Chris Uren

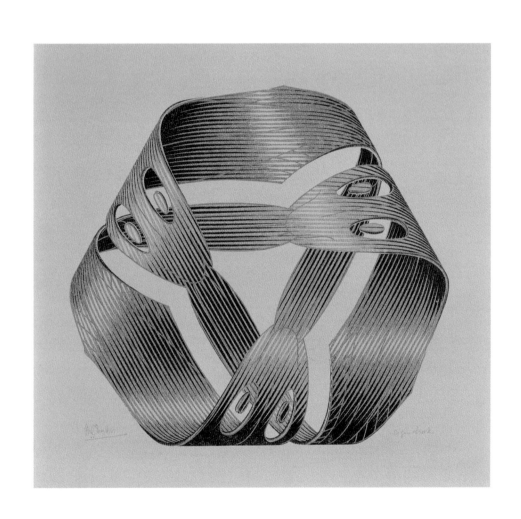

Knots Unravelled

From string to mathematics

Meike Akveld

and

Andrew Jobbings

Shipley, United Kingdom

Knots Unravelled

Published by Arbelos.

PO Box 203, Shipley, BD17 5WT, United Kingdom
http://www.arbelos.co.uk

First published 2011.

Cover illustration and typographic design by Andrew Jobbings.
Typeset with LATEX.

Printed in the UK for Arbelos by The Charlesworth Group, Wakefield.
http://www.charlesworth.com

ISBN 978-0-9555477-2-0

Contents

Buddhism, and various knots used as Christian symbols (see figure 1.1). We shall learn later that the *trinity knot* is mathematically speaking a 'link', rather than a knot, because more than one strand is involved.

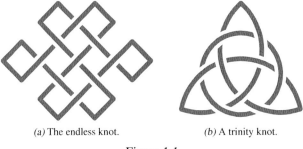

(a) The endless knot.   (b) A trinity knot.

— *Figure 1.1* —

More recently, knots have featured in logos; perhaps unsurprisingly they are especially popular among mathematicians. Figure 1.2 shows the logo of the IMU, the International Mathematical Union (though the actual logo is in colour [20]), and a logo representing the Cantons of Switzerland, designed by Sebastian Baader for the Swiss Knots 2009 conference [26]. Once again, we shall see later that both of these are actually 'links' rather than knots.

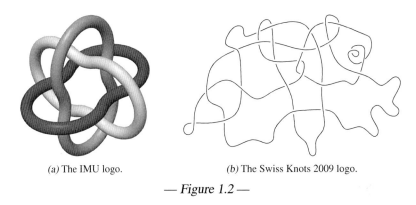

(a) The IMU logo.   (b) The Swiss Knots 2009 logo.

— *Figure 1.2* —

Some modern sculptors have been inspired by mathematical knots, notably John Robinson (some of whose works are in display at the University of Wales, Bangor [12]), Alex Feingold [15], Helaman Ferguson [16] and Carlo Séquin [24].

Knots also feature among the works of the well-known Dutch graphic artist M. C. Escher, one of which shows three trefoil knots (the trefoil is discussed on page 23).

In contrast to these representations of knots in art, knots also play an important role in modern science, occasionally in unexpected ways. For example, cell biologists have discovered that DNA deals with potentially harmful knots by using 'unknotting' enzymes called topoisomerases [33]. Knot theory has also been applied in fields ranging from polymer chemistry to quantum gravity.

1.2 Knots in rope

What a rope-user usually wants to know is the answer to the question "What is the best knot for this purpose?". To this end, knot-tiers classify knots into broad categories that reflect the use of a knot.

In rope-users' terminology, the only way to produce a 'true' knot is to tie a single rope to itself. Examples of 'true' knots include stopper knots such as the figure of eight, binding knots such as the reef knot, and loops such as the bowline. Strictly speaking all other cases are not knots: these include bends, which join two ropes, and hitches, which attach a rope to another object.

As we shall see, mathematicians use the term 'knot' with a rather different meaning, and also have a different method of classifying knots.

1.3 Knots from an engineering perspective

In the core of this book we study knots from a mathematical perspective, whereby a knot is perceived in an abstract way. Knots may be studied from an engineering or scientific perspective, treating them as 'real' objects.

For example, all knots reduce the breaking strength of rope: a single knot can reduce the strength by up to 50%. It is possible to determine the factors affecting the strength of a knot, and to measure how much a rope has been weakened by tying a knot in it. The actual material from which the rope is made is clearly important, but the strength of a knot is also influenced by the thickness and the shape of the cross-section of the rope.

But strength is not the only important measure of a knot's usefulness. Security (whether the knot holds firm under various adverse conditions) and releasability (how easy it is to untie the knot) are also important.

It turns out that a key role is played in all these measures by the geometry of the knot: the way in which the rope crosses itself; how tightly the rope bends. And some aspects of the geometry of knots form the basis of mathematical knot theory. So the abstract mathematics proves to be important even when looking at knots from the 'real' perspective.

1.4 History of the mathematics of knots

The first mathematical investigations of knots were in the eighteenth century. In a 1771 paper, the French mathematician Alexandre-Théophile Vandermonde briefly refers to knots. Though the paper deals with the problem of the knight's tour on the chess board, Vandermonde considers the intertwining of the curves generated by the moving knight. The great mathematician Carl Friedrich Gauss (1777–1855) almost certainly thought about knots, but published nothing on the subject. It was a Czech student of Gauss, Johann Benedict Listing, who really began the mathematical study of knots. Listing's 1847 book *Vorstudien zur Topologie* contains many early topological ideas, including the problem of differentiating knots, but his work was largely ignored.

The first systematic approach began in 1876, when the Scottish physicist Peter Guthrie Tait started to classify knots according to how many crossings they had. His aim was to identify the so-called prime knots, which can be used to construct any other knot, just as any number can be obtained by multiplying together prime numbers. Tait was later assisted by America's first knot theorist C. N. Little.

This work was motivated by a theory of William Thomson (Lord Kelvin), in which atoms were vortex rings, whose different shapes would account for the different chemical elements (see figure 1.3). The theory was later considered mistaken, and physicists lost interest in knots until more recent times.

From Thomson's "On Vortex Motion".
— *Figure 1.3* —

Tait's work revealed a general feature of knot theory: even the simplest questions can be surprisingly difficult to answer. It took until the 1920s before mathematicians proved the 'obvious' fact that knots can exist! As we shall see, this is not as ridiculous as it sounds—it is remarkably difficult to decide whether a tangled piece of string is knotted.

We shall also learn how difficult it can be to tell whether two knots are the same or different. One historical episode illustrates this very well: in 1974 the lawyer Kenneth Perko discovered that two diagrams in Tait and Little's table actually represent the same knot. The knots, now known as the *Perko pair*, are shown in figure 1.4.

— *Figure 1.4* —

Intuition is not really much help here—these two knots certainly *appear* to be different. The aim of this book is to show how mathematics can help to deal with such problems.

Knots in paper

A regular pentagon

One way to make a regular polygon is to tie a knot in a strip of paper.

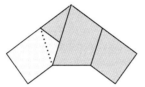

> **Activity 1** Take a strip of paper and tie a simple 'overhand' knot in it. Carefully pull the strip tight and flatten the knot. Fold over (or cut off) the ends of the strip. Confirm that the result is a regular pentagon.

The Möbius band

Take a strip of paper and bring the ends together. Joining the ends at this stage would just make a cylinder. Instead, give one end of the strip a single half-turn before joining the ends together. The result is now a Möbius band, sometimes called a Möbius strip, which is a *one-sided surface*: it is possible to connect any two points on the surface by drawing a line which does not cross the boundary.

Activity 2   What happens when you cut a Möbius band 'in thirds' by cutting all the way along a line one third of the way in from the edge? (The line is shown dashed in the figure on the preceding page.)

Activity 3   Suppose you put *three* half-twists in the strip before joining the ends together. What happens when you cut the resulting band 'in half' by cutting all the way along the centre line?

Möbius shorts

Take a T-shaped piece of paper, as shown in the upper-left figure below. Glue the ends X and Y together to make a ring (without a twist). Now pass the end Z up through the ring and bend the strip over to where X and Y meet, as shown in the main figure. Finally, glue Z to the ring to make the Möbius shorts, which form another one-sided surface. The surface was given its English name by Ralph Boas [29].

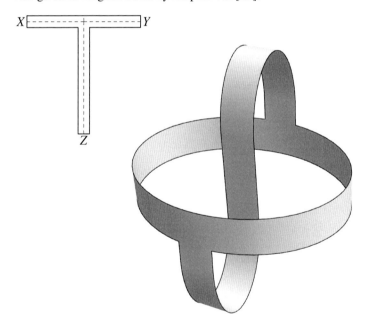

Activity 4   What happens when you cut the Möbius shorts 'in half', by cutting the ring and the connecting stem all the way along their mid-lines (shown dashed in the upper-left figure)?

CHAPTER 2

Working with diagrams

2.1 Describing knots

> **Task 2.1** Figure 2.1 shows two knots. Make these knots with your own rope (for the second knot you need two ropes).
>
> Do you know the name of these knots and where they are used?

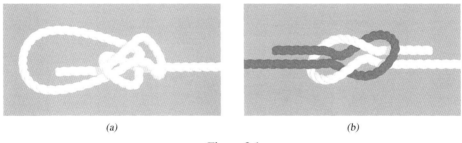

(a)   (b)

— Figure 2.1 —

Task 2.1 shows that it is not hard to tie a given knot provided you have a good picture. In particular it is important that you are able to see very clearly whether a strand of the rope passes over or under another strand. For this reason, when you make a knot for study purposes you should never pull the knot together tightly, but leave the strands loose so that you can see the crossings clearly. Of course, when you are actually using a knot for real you will want to pull the strands tight!

> **Task 2.2**　Make your favourite knot with your piece of rope. Now imagine describing this knot over the phone to a friend, so that they can also make the knot. (No cheating—you are not allowed to use a phone camera!)

You almost certainly found that task 2.2 was much harder than task 2.1. It is not so easy to describe a knot in words.

We obviously need a clear way to represent a knot graphically; it would be a bonus if we could describe a knot without a picture, perhaps with a formula. In any case we will need a systematic approach. We shall first consider how to draw accurate diagrams, but before we do so we need to explain more precisely what we mean by a knot in mathematics.

2.2　Mathematical knots

The knots in figure 2.1 are not what a mathematician thinks of as a knot—they have ends! It is simpler to ignore the ends and think of the rope as being one continuous loop.

The best way to explain this is to consider a knot in a rope. So take a single piece of rope and tie it together in any way you like. Now here is the difficult part: you have to bring the two ends of the rope together in such a way that it is not possible to see where the rope was interrupted. Of course this is almost impossible to do in practice, but you can no doubt imagine it—something like a knotted rubber band (see figure 2.2). This is what mathematicians mean by a knot. (Actually, this is not true—a mathematician also requires the 'rope' to have no thickness, like an idealised circle in geometry.)

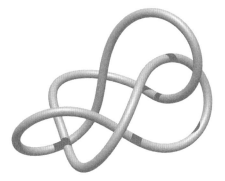

A mathematical knot.

— Figure 2.2 —

Perhaps you did not follow our advice to use a piece of rope. You possibly thought that it would be sufficient to use your imagination (indeed, so far that was not too hard). And there is a sense in which you are right; generally speaking, imagining is a very important ability for a mathematician.

However, for this book you really need a piece of rope. Make sure it is long enough and thick enough, so that it is easy to tie knots in it. To bring the two ends together it is probably simplest to use a safety pin; then you can also easily open the knot again.

2.3 Projections and knot diagrams

One way to represent a knot is with a *knot projection*: we project the knot on a plane. Or, less mathematically but more concretely, we consider the shadow of a knot on the ground. Figure 2.3 shows what happens when the knot in figure 2.2 is projected.

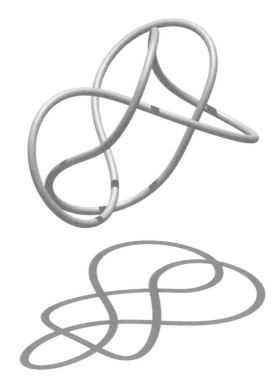

Projecting a knot on a plane.

— *Figure 2.3* —

Task 2.3 Make a (mathematical) knot in your rope—it should not be too complicated. Now make a sketch of the real knot and its projection.

If you have difficulty doing this, then you can use your desk light. Put a piece of white paper on your desk and hold the knot under the light above the paper. You can now simply trace the shadow of the knot on the paper.

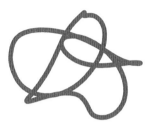

A poor projection.

— *Figure 2.4* —

When projecting the knot we have to be careful that the resulting figure is clear and unambiguous, unlike figure 2.4, for example. To do this we ensure that each crossing is of the simplest possible form—two strands crossing more or less at right angles—and avoid some rather special situations (see figure 2.5):

 ❧ There are no "cusps", where a strand doubles back on itself.

 ❧ Two strands of rope are not allowed to touch each other.

 ❧ Only two ropes should intersect at a crossing, not three or more.

(a) Admissible.   *(b)* Not admissible.

Admissible and inadmissible crossings.

— *Figure 2.5* —

When this is achieved, as in figure 2.6, mathematicians speak of a *generic projection*. To insist on a generic projection is no great limitation. The knot does not change if you move it around, as a whole, in space, so you can always turn it in such a way that non-admissible situations do not occur.

A generic projection.

— *Figure 2.6* —

The problem with a knot projection is that key information about the knot is missing: at the crossings, you do not see which strand is above and which below. This information is vital for a full description of the knot. To obtain a useful diagram we need to include information about the nature of the crossings; the result is known as a *knot diagram*.

To form a knot diagram, take a projection of the knot and indicate at each crossing which strand passes below, and which above, by drawing an interrupted line for the lower strand. Figure 2.7 shows the result of doing this for the knot in figure 2.2 on page 8.

A knot diagram.

— *Figure 2.7* —

Task 2.4 By redrawing the crossings, change the knot projection you drew for task 2.3 into a knot diagram.

We shall use knot diagrams extensively throughout the remainder of this book.

2.4 Knotted or not? The same or different?

Task 2.5 Consider the knot diagrams in figure 2.8. Determine first whether these diagrams really represent knots—are they truly knotted?

For those which are knotted, decide whether the diagrams show different knots, or do two (or more) actually show the same knot?

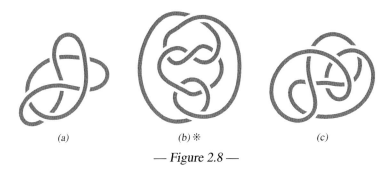

(a)   (b) ✳   (c)

— *Figure 2.8* —

Task 2.5 raises two of the central questions of knot theory:

- When can a knot be deformed into an unknotted piece of string?
- How can we decide whether one knot can be deformed into another?

By "deform" we mean that we are allowed to play with the knot as much as we like—we can move strands of rope wherever and however we like—but it is forbidden to use a pair of scissors.

The unknot

> **Task 2.6**   Use your own rope to make the knots shown in figure 2.9. How simple can you make each knot by deforming it?

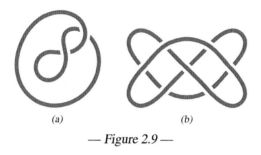

(a)   (b)

— *Figure 2.9* —

Both the knots shown in figure 2.9 can be deformed into a circular ring—the simplest shape that a closed loop of rope can form. Mathematicians call a knot which can be deformed into a circular ring the *unknot*.

> **Task 2.7**   Decide whether each of the knot diagrams in figure 2.10 represents the unknot.

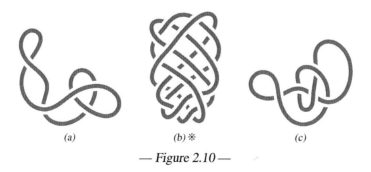

(a)   (b) ✳   (c)

— *Figure 2.10* —

Note in passing that the situation here is typical in mathematics, where the trivial or the neutral is given its own name and can be quite important. Think of adding two numbers:

adding 0 to a number leaves the number unchanged, so has no effect, but the concept of zero is an important one, taking many centuries to develop and even longer to be accepted.

Equivalent knots

When two knots can be deformed into each other we call them *equivalent* knots or simply the *same* knot.

> ✱ **Task 2.8**   Determine whether the knot diagrams in figure 2.11 represent two different knots or equivalent knots.

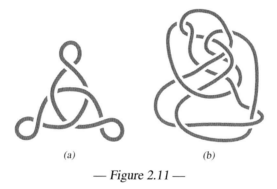

(a)   (b)

— *Figure 2.11* —

You have probably realised that it is not so easy to determine whether or not two knots are equivalent. Of course, you can simply try for a while with piece of rope. If you manage to deform one knot into the other you are done. But what if you do not manage? Are the knots not equivalent, or did you give up too early, or were you simply unlucky?

2.5   Reidemeister moves

When we try to deform a knot in order to simplify it, we move the strands of the rope around. We do this more or less instinctively, apparently understanding intuitively which moves are sensible and which are not. In the process a projection of the knot will change and hence the corresponding knot diagram will also change—we may create new crossings or destroy old ones.

In the 1920s the German knot theorist Kurt Reidemeister (1893–1971) investigated this process systematically [32]. Surprisingly, he discovered that if you have knot diagrams of two equivalent knots, then you can change from one to the other by using only three types of moves, the *Reidemeister moves* (see figure 2.12 on the following page). It often happens in mathematics that the same result is discovered independently by more than one person; in this case, Reidemeister's result was also discovered, around the same time, by the American mathematicians James Alexander and G. B. Briggs [28].

Each move is considered to take place within a distinct region without affecting any other part of the knot.

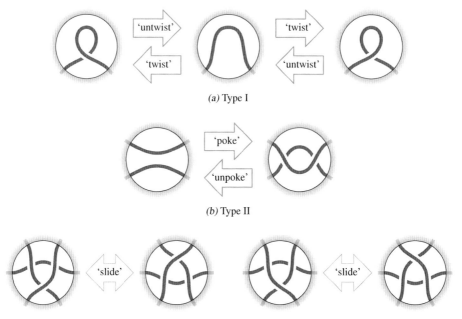

(a) Type I

(b) Type II

(c) Type III (two versions)

The three Reidemeister moves.

— *Figure 2.12* —

Type I A move of type I allows us to put in or remove a twist in a single strand (see figure 2.12a).

Type II A type II move applies to two strands, allowing us to add or remove two crossings by moving a loop of one strand across another (figure 2.12b).

Type III A move of type III applies to three strands and lets us slide a strand from one side of a crossing to the other side (figure 2.12c).

For a move of type III, it doesn't matter whether you think of the top, middle or bottom strand (taking the diagram to be looking down from above) as the one being moved. After the slide, in theory you need to nudge the three new crossings towards the centre to obtain the given figure, but in practice we just think of sliding one strand to the other side of the crossing point of the other two.

Note that there are two versions of type III because there are two possible (mirror-image) orientations of the strands. Also, the reverse of a move of type III is actually just the same as the move itself, rotated through 180°.

It should be reasonably clear that these three moves do not change the knot. What is perhaps more surprising is that these three moves suffice, so that we do not need more moves, or more complicated ones. Though a proof of Reidemeister's result is not very hard, it is beyond the scope of this book.

Task 2.9 In figure 2.13, the top diagram can be changed into each of the other diagrams using a single Reidemeister move. Determine which move, and at which crossing(s) it needs to be performed.

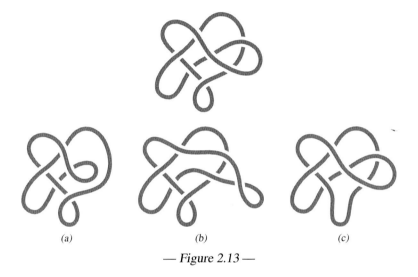

(a) (b) (c)

— *Figure 2.13* —

Task 2.10 You discovered in task 2.6 on page 12 that the knot diagram in figure 2.9b shows the unknot. Verify Reidemeister's result for this example, by writing down precisely which moves change figure 2.14a into figure 2.14b.

There is more than one way to do this. Can you find a method using as few moves as possible?

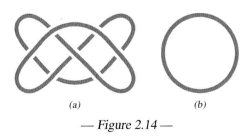

(a) (b)

— *Figure 2.14* —

Task 2.11   Suppose you perform just one Reidemeister move on the knot diagram in figure 2.15. Which of the moves can you use?

— *Figure 2.15* —

Celtic knots

There is a long tradition of abstract geometric designs in the art of the Celtic peoples of ancient Britain and Ireland. Complex knotwork patterns were used profusely in Celtic illuminated manuscripts, two of the best known being the Book of Kells (middle 6th to early 8th century, now in the library at Trinity College, Dublin) and the Lindisfarne Gospels (late 7th century, now housed in the British Library). In these manuscripts interlaced designs are used both to fill areas and as borders for text and illustrations. Such a design, where each strand crosses alternately over and under, is called a *Celtic knot*, for example:

Interlaced designs of this type also feature on Celtic crosses—monumental stone crosses found mainly in the British Isles (especially Ireland, Scotland, Wales and Cornwall), which were erected from as early as the 7th century until the mid 12th century.

Celtic knots are similar to designs found in other cultures. Some Roman mosaics feature simple interlaced designs, and interlace patterns are common in Islamic design. Some simpler Celtic knots are also related to so-called mirror curves, such as those found in *sona* or *kolam drawings* (see [22], for example).

Plaited rectangles

The simplest form of Celtic knot is the *plaited rectangle*, where the strands form a simple 'over and under' interlacing pattern. The left-hand figure on the next page shows 3×5 and 3×9 plaited rectangles: in a $p \times q$ plaited rectangle the strands meet one side of the rectangle p times and an adjacent side q times.

More complicated Celtic knots can be obtained from a plaited rectangle by 'removing' some crossings. To remove a crossing, replace the crossed pair of strands by an uncrossed pair (there are two ways to do this: see figure 7.2 on page 78). The right-hand figure below shows a Celtic knot obtained in this way from a 3 × 5 plaited rectangle.

Plaited rectangles.   Two crossings removed.

Activity 1   Determine which crossings of a 3 × 9 plaited rectangle need to be removed in order to obtain each of the following Celtic knots.

Not all Celtic knots are rectangular, however, as the examples on the previous page illustrate. Unfortunately, there is no record of the methods used by the monks and craftsmen who created the original Celtic knots.

How many strands?

It is not always easy to see at once how many separate pieces of rope are involved in a given Celtic knot. (We shall see later that, in mathematics, a knot using more than one rope is strictly called a link, and that the separate strands of a link are referred to as the components—see chapter 6.)

Activity 2   Find the number of strands in each of the Celtic knots shown above and on the previous page.

For simple plaited rectangles it is possible to determine the number of strands directly from the dimensions of the rectangle, using the result: in a $p \times q$ plaited rectangle, the number of strands is the highest common factor of p and q. A proof of the result may be found in [30].

Activity 3   Check the result for the 3 × 5 and 3 × 9 plaited rectangles in the figure above.

Use the result to find the number of strands in a 4 × 10 plaited rectangle.

At the time of writing, there is no known way to determine the number of strands in a general Celtic knot, apart from drawing and counting. Note that changing a crossing may increase or decrease the number of strands, or leave it the same.

CHAPTER 3

Counting crossings

3.1 Telling knots apart

Does figure 3.1 show six different knots, or are some of them the same? How does one distinguish between knots, that is, determine whether they are the same or different? This is the central question of knot theory and this chapter discusses a partial answer.

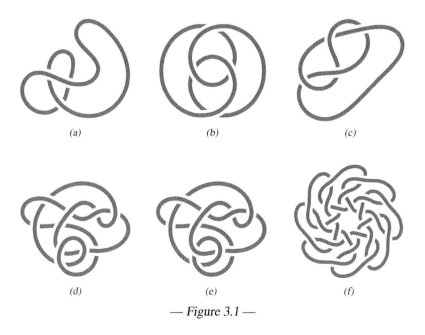

(a)   *(b)*   *(c)*

(d)   *(e)*   *(f)*

— Figure 3.1 —

3.2   The crossing number

When a diagram of a knot is drawn, it is an easy matter to count the number of crossings in the diagram. Is this number a useful property of the knot? Will it help us to distinguish between two knots?

Consider the two knots shown in figure 3.2. The left-hand knot has three crossings, the right-hand knot has four crossings. But if you make the knots and move them around, or try to visualise this process, you will see that these are actually the same knot: in fact they are both the unknot. So the same knot can have different numbers of crossings, depending on how the knot diagram is drawn.

Knot diagrams with different numbers of crossings.

— *Figure 3.2* —

Does this only happen with the unknot? No, the same difficulty arises in more complicated knots, such as those in figure 3.3.

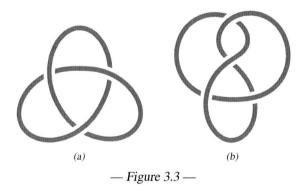

(a)   (b)

— *Figure 3.3* —

Task 3.1   How many crossings are there in each of the knot diagrams in figure 3.3?

Can the right-hand knot be transformed into the left-hand knot?

We see that the *same* knot can have *different* numbers of crossings. The situation is actually much worse: *different* knots can have the *same* number of crossings. Consider the knot diagrams in figure 3.4.

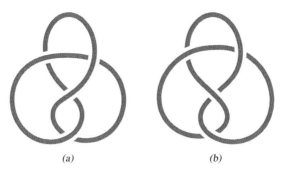

(a)   (b)

Knot diagrams with the same number of crossings.

— *Figure 3.4* —

The two knot diagrams in figure 3.4 have the same number of crossings. What can we say about the knots? Well, we met the left-hand knot (rotated through 180°) in figure 3.3b on the facing page. Experimenting with the right-hand knot should convince you that it is a different knot. In fact, you will find that you cannot reduce the number of crossings below four. This means that we know we have two different knots—there is a diagram of the left-hand knot with only three crossings (figure 3.3a), whereas the right-hand knot has no such diagram.

We learn that, rather than just counting the number of crossings in a given diagram, the really useful number to find is the *minimal* number of crossings among *all* possible diagrams. This minimum number is called the *crossing number* of a knot. From the way it is defined, it follows that the crossing number is a *knot invariant*—equivalent knots are always assigned the same value (although different knots may have the same knot invariant).

> ✳ **Task 3.2**   Determine the crossing number of each of the knots in figure 3.1 on page 19. You may find it helpful to make each knot from a piece of string, so that you can establish which crossings can be removed and which cannot.

3.3   Which crossing numbers are possible?

In section 3.1 we saw that every knot has a crossing number. What happens when we ask the question the other way round: is every natural number the crossing number of a knot? Let us consider the first few cases in turn.

Crossing number 0

We have already met a knot with crossing number zero, which is the crossing number of the unknot (and clearly any knot with no crossings is the unknot).

Crossing numbers 1 and 2

Task 3.3   Can you find a knot with crossing number 1 or 2?

Although you may easily discover knot diagrams with only one or two crossings, such as those in figure 3.5, you should find that the resulting knot is always the unknot. The crossing number of such a knot is therefore always zero.

Knot diagrams with 1 or 2 crossings.

— Figure 3.5 —

It is possible to show that there are no knots with crossing number 1 and 2. For example, suppose that a knot has exactly one crossing, as shown in figure 3.6.

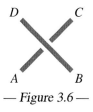

— Figure 3.6 —

Consider the four ends emanating from the crossing. These have to be joined up in pairs, without forming any other crossings. There are only four ways in which this can be done, the two diagrams shown on the left of figure 3.5 and two others which are rotations and reflections of these. For example, joining A to B and C to D gives the left-hand knot diagram. All of the resulting four knots are the unknot, and so have crossing number 0.

❋ **Task 3.4**   Show that any knot with exactly two crossings is the unknot.

Crossing number 3

We met one knot with crossing number 3 in figure 3.3, shown again in figure 3.7a. Are there any others? One possible candidate is shown in figure 3.7b.

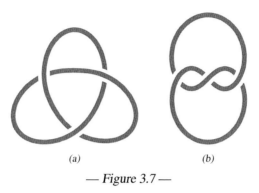

(a)   (b)

— Figure 3.7 —

Task 3.5   Find the crossing number of the knot in figure 3.7b.

Are the two knots in figure 3.7 different?

The knot in figure 3.7a is called the *trefoil knot* (also see figure 3.8a). The word 'trefoil' literally means 'three leaves', from the Latin *trifolium*. The knot in figure 3.7b is often called an *overhand knot* (also see figure 3.8b); it is what you might use when you tie up a parcel or tie your shoe laces.

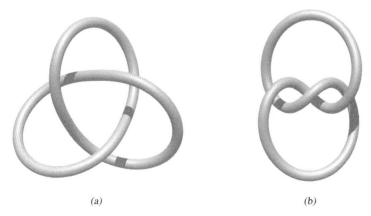

(a)   (b)

The trefoil and overhand knots.

— Figure 3.8 —

The overhand knot and the trefoil knot are the same; in fact, it turns out that the trefoil is the *only* knot with crossing number 3. Therefore, in this sense, the trefoil is the simplest possible true knot. In other words, the trefoil is the knot with the smallest possible crossing number without being the unknot.

> **Task 3.6**   Convince yourself that the trefoil and overhand knots are the same by changing the knot in figure 3.7b into the knot in figure 3.7a

Showing that there is only one knot with crossing number 3 is rather involved, but we can give an outline of one possible method. We start by considering knot projections with three crossings. There are essentially only four possibilities, those shown in figure 3.9.

Knot projections with 3 crossings.

— *Figure 3.9* —

Now consider modifying the crossings in each of these projections to form a knot diagram. The first two projections in figure 3.9 can be eliminated, because, whatever form the crossings take, the loops can be untwisted, thus reducing the number of crossings and eventually leading to the unknot.

We are left with the last two projections in figure 3.9. When the crossings are modified, the first of these leads to a knot diagram of either the unknot or the trefoil knot; the second leads to the unkot or the overhand knot, but we know that the overhand knot is the same as the trefoil.

> **Task 3.7**   Convince yourself that the knot projections shown in figure 3.9 are essentially the only possible ones with 3 crossings.

Crossing number 4

We have already met a knot with crossing number 4, the knot shown in figure 3.4b on page 21, and it turns out that there are no other knots with four crossings. Known as the *figure of eight knot* and illustrated in figure 3.10 on the next page, the knot is more fully described in the interlude on page 47.

The figure of eight knot.

— *Figure 3.10* —

3.4 Does the crossing number classify knots?

Table 3.1 summarises the results we have found so far. The table shows all the knots corresponding to each crossing number from 0 to 4.

Crossing number	Knot
0	unknot
1	–
2	–
3	trefoil
4	figure of eight

— *Table 3.1* —

Task 3.8   Use table 3.1 to determine whether each of the knots in parts (a) to (e) of figure 3.1 on page 19 is the unknot, the trefoil or the figure of eight.

We may be tempted to think that table 3.1 can be continued to provide a convenient way to tell all knots apart. Does the crossing number provide a simple means of classifying knots completely? Unfortunately, the answer is 'no': the strategy fails since there are *different* knots with the *same* crossing number, as we will see when we consider knots with crossing number 5. Indeed there are at least two different knots for *every* crossing number greater than 4.

All is not lost, however. We do know that knots with *different* crossing numbers are definitely not the same. What we cannot do, without investigating further, is make any deduction about two knots with the same crossing number, unless the crossing number is 0, 3 or 4 when table 3.1 can be used.

Despite this, when tables of knots are produced the knots are usually categorised by the crossing number.* But for crossing numbers greater than four more than one knot is listed in the table.

There is another drawback: calculating the crossing number is not as straightforward as it seems. The examples we have considered have not been very complicated, but even for such simple knot diagrams it is not always easy to see which crossings can be removed and which cannot. Now imagine trying to do this for a knot diagram with a very large number of crossings, such as that shown in figure 3.11.

— Figure 3.11 —

3.5   Crossing number 5

Task 3.9   Convince yourself that the two knots in figure 3.12 on the next page have crossing number 5.

The knot in figure 3.12a has a similar structure to the trefoil but has five lobes; it is known as the *cinquefoil knot*. The knot in figure 3.12b is known as the *3-twist knot*. But perhaps the two diagrams are just rearrangements of the same knot?

Task 3.10   Are the two knots in figure 3.12 the same or different? Make one of the knots and see if you can turn it into the other knot.

After a little experimentation you should get the strong impression that the knots are different. However, when we want to distinguish between more complicated knots we will need better methods than simply trial and error. We consider one possible method next.

*This was in fact the method adopted for the first ever table of knots, published by Peter Guthrie Tait in 1877.

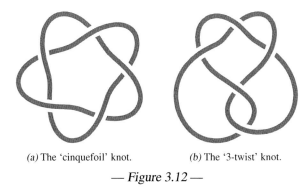

(a) The 'cinquefoil' knot.　　　　　*(b)* The '3-twist' knot.

— *Figure 3.12* —

Words from labelled crossings

With any knot diagram we can associate a "word", obtained by labelling all the crossings and then traversing all the way around the diagram, as follows.

1. Label all the crossings in the knot arbitrarily with the letters A, B, C, ... (So for five crossings we use the letters A to E.)
2. Choose a starting point on the knot.
3. Walk along the knot from the starting point and note down every letter that you pass.
4. Keep going until you get back to your starting point.
5. The sequence of letters forms a "word" associated with the knot diagram.

Note that in this process you pass every crossing twice. Can you see why?

> **Task 3.11**　Find a word associated with each of the two knot diagrams in figure 3.12. Compare the two words.

You probably noticed that the knot diagram in figure 3.12a yields a word which repeats a sequence of five letters twice, whereas for figure 3.12b the second five letters seem to be a random mix of the first five.

> **Task 3.12**　Consider the word associated with the 3-twist knot which you found in task 3.11. Organise your letters in such a way that the word starts with 'ABCDE'. Find a rule that describes the order of the remaining letters.

It turns out that any knot diagram of the cinquefoil knot yields a word that repeats a five-letter sequence, whereas any word associated with the 3-twist knot never has this property. We can therefore use this method to tell these two knots apart.

> **Task 3.13** Use the method on the preceding page to determine which of the knot diagrams in figure 3.13 represent the cinquefoil and which represent the 3-twist knot.

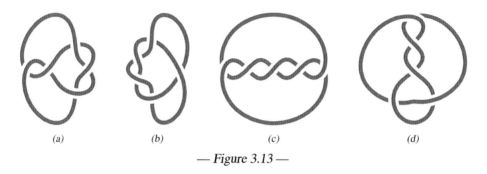

(a)   (b)   (c)   (d)

— Figure 3.13 —

We have now seen that there really are two different knots with crossing number 5. But is this all, or are there more knots with this crossing number? For the time being we leave this question open; in order to give the answer we have to wait until we have developed a new technique.

3.6 Classifying knots

The fact that there are two knots with crossing number 5 has an important consequence: knot theory is actually much more complicated, but also much more interesting, than may have appeared before we considered crossing number 5. Before then it looked as though determining the crossing number was sufficient to classify a knot and, although this may be quite hard for a knot with many crossings, that could have been the end of knot theory. Luckily this is not the case!

The bigger the crossing number, the more knots we find with that particular number of crossings. In the table on page 111 you will see a list of all the knots up to crossing number 8. For 5 crossings there are only two knots, but for 8 there are already twenty-one. Indeed, the number of knots belonging to a certain crossing number actually grows exponentially—there are 9988 knots with 13 crossings and 1 388 705 with 16 (see [25]; these numbers are for prime knots, defined later in figure 4.10 on page 39).

Nevertheless, we do seem to have a strategy for distinguishing between two knots. Firstly, determine the crossing number of each knot. If these numbers are different, then we are finished—the knots are not the same. However, if the crossing numbers are equal, we need a method to determine whether or not they are the same.

Unfortunately, the method of the previous section—finding a word associated with the knot diagram—is not suitable in general: since the method takes no account of the nature of each crossing, changing an over-crossing to an under-crossing leaves the associated word unchanged. So there are many examples of different knots which give the same word, even knots with the same crossing number, such as those in figure 3.14.

Two different knots giving the same associated word.

— Figure 3.14 —

> ✳ **Task 3.14**   Use the method of section 3.5 to show that the two knot diagrams in figure 3.14 lead to the same words. Convince yourself that the knots are actually different.

So we need new methods to distinguish between knots with the same crossing number. Furthermore, given the difficulties involved in finding the crossing number, it would be a bonus if the methods worked for any two knots, thus avoiding the need to find crossing numbers at all. This is the real goal of the remainder of this book, but first we need to be a little more precise about the nature of the knots we are dealing with. There are some important issues we have not discussed so far, such as:

    𖠰 Mirror images—are a knot and its reflection the same?

    𖠰 Combining knots—is it possible to break a knot down into simpler component parts?

These are the themes of the next chapter.

INTERLUDE

Tie knots

Anyone who regularly wears a tie soon learns that there is more than one type of knot that can be used. One of these is the Windsor knot.

The Windsor knot

The Windsor knot, supposedly named after the Duke of Windsor, is often used when a wide knot is required in a tie. The figures show how it is tied (when looking in a mirror). At Step 9 the front of the tie has been tucked through the last loop of the knot.

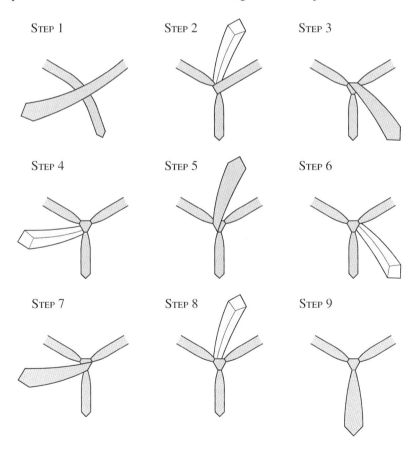

> **Activity 1** Make a Windsor knot in a tie and remove the tie by slipping it over your head, keeping the knot intact. Now hold the two ends of the tie together, as if they were joined, so that the tie forms a knotted loop. (You may wish to use a safety pin to keep the ends together.)
>
> The resulting knot is one we have already met in this book—which one?

Classifying tie knots

In 1989 the first new tie knot to appear for 50 years was widely publicised. The knot is known as the *Pratt knot* after its inventor, Jerry Pratt, who had been wearing the knot for about twenty years before it was popularised by a television presenter. The Pratt knot is unusual in that the tie starts off 'reverse side out' when the knot is tied.

Only four tie knots are traditionally used: the 'four-in-hand' (the basic knot learnt by schoolchildren), the Nicky (a derivative of the Pratt knot), the half-Windsor (a simplified version of the Windsor) and the Windsor. Descriptions of these and other knots can be found on many websites, including [17].

Once a new knot has been discovered, a natural question arises, are there any others?

Not only have new knots been found, they have *all* been found—there is a complete classification. In consequence no new tie knots can be found in future.

In 2000 Thomas Fink and Yong Mao published an article in which they analysed tie knots mathematically. In the process they calculated and classified all possible tie knots, showing that there are exactly 85 possible ways in which a conventional tie can be tied. They also discovered nine 'aesthetic' knots not in normal use. For more information see the book [5] or the websites [17, 18].

CHAPTER 4

New knots from old

4.1 Mirror images

Suppose you make a trefoil with your rope and hold it in front of a mirror. What do you see? You will see another trefoil but with reversed crossings: what used to be in front is now at the back and vice versa, see figure 4.1.

The trefoil knot and its mirror image.

— *Figure 4.1* —

We can obtain a knot diagram of the mirror image of a knot by reflecting a diagram of the original knot in a mirror placed behind the diagram, parallel to the plane of the paper. In general, reversing all the crossings in the original knot diagram—changing over-crossings to under-crossings and *vice versa*—has the same effect. The result of doing this for the trefoil is shown in figure 4.2.

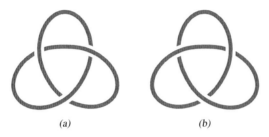

(a)   (b)

Knot diagrams of the trefoil and its mirror image.

— *Figure 4.2* —

Now we stated on page 24 that there is only one knot with crossing number 3. In that case, in figure 4.1 the knot in front of the mirror and its reflection should be the same knot: we should be able to transform the knot into its mirror image

> **Task 4.1**   Make the trefoil with one rope and its mirror image with another. Can you deform one knot into the other?

You probably discovered in task 4.1 that, no matter how long you try, you will not succeed in deforming the trefoil into its mirror image. But we now know that this is not sufficient to prove that they are different—perhaps we were simply unlucky. Later in the book we will prove that the trefoil and its mirror image are indeed different knots, but in order to do this we need some more techniques. The first proof was given by the German mathematician Max Dehn in 1914.

A knot which is *not* equivalent to its mirror image, such as the trefoil, is called *chiral*. In practice it is not easy to distinguish the trefoil from its reflection. Just as we speak of left-handed and right-handed threads, we could speak of a left-handed and a right-handed knot. But even if we did so, we would have to decide which was which.

> **Task 4.2**   Decide whether each of the diagrams in figure 4.3 on the facing page represents the trefoil knot (figure 4.2a) or its mirror image (figure 4.2b).

> **Task 4.3**   Make the figure of eight knot (figure 3.10 on page 25) and its mirror image out of two pieces of rope. Try to deform one into the other.

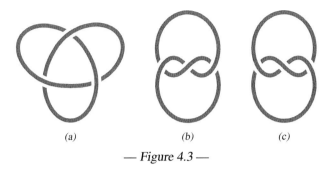

(a)   *(b)*   *(c)*

— *Figure 4.3* —

✳ **Task 4.4**   Describe a sequence of Reidemeister moves changing the knot diagram of the figure of eight knot (figure 4.4a) into its mirror image.

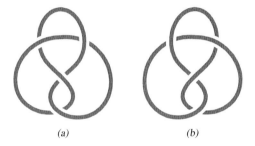

(a)   *(b)*

Knot diagrams of the figure of eight and its mirror image.

— *Figure 4.4* —

When a knot is the *same* as its mirror image we call the knot *amphichiral* (the term *achiral* is sometimes used, or the spelling *amphicheiral*). We conclude from task 4.3, or task 4.4, that the figure of eight knot is amphichiral.

✳ **Task 4.5**   Consider the two knots with crossing number 5: the cinquefoil and the 3-twist (see figure 3.12 on page 27). In each case, decide whether the knot is chiral or amphichiral.

In chapter 7 we will meet a method which helps to decide whether a knot and its mirror image are equivalent.

For a long time all the known amphichiral knots had even crossing numbers. As early as 1890 Peter Guthrie Tait conjectured that any knot with an odd crossing number was different from its mirror image. However, in 1998 an amphichiral knot with crossing number 15 was discovered (see figure 4.5 on the next page) and it was also verified that there is no such knot with a smaller odd crossing number. Furthermore, in 2008 it was proved that there is an amphichiral knot for every odd crossing number greater than 15.

An amphichiral knot with odd crossing number.

— Figure 4.5 —

When classifying knots, mathematicians usually consider a knot and its mirror image as forms of the *same* knot. From this viewpoint, the chirality of the knot (whether it is chiral or amphichiral) is just another property of the knot.

One consequence of this approach is that mirror images are not normally included in tables of knots. We adopt the same convention in the table on page 111, where mirror images are not given, even when they are in fact different from the knot shown.

With this convention, the statement we made on page 24 is true—there is only one knot with crossing number 3, the trefoil. But the trefoil is chiral, so we consider it as having left-handed and right-handed forms.

> **Task 4.6**   By checking the crossings, confirm that the two knot diagrams in figure 4.6 are mirror images.
>
> What happens when you rotate the left-hand diagram through 90° clockwise? Explain why this shows that the knot is amphichiral.
>
> Make the knot using a piece of rope and see whether it is one you recognise.

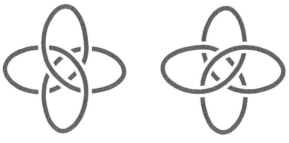

— Figure 4.6 —

Task 4.7 What happens when you rotate the knot diagram in figure 4.7 through 180°?

Is the knot chiral or amphichiral?

— Figure 4.7 —

Task 4.8 Suppose you make a knot from a piece of rope and place it on a table. When you lift the knot from the table, turn it over, and put it down again, you do not (in general) get a mirror image. The knot has not changed, so you get the same knot again.

However, when you draw the corresponding knot diagram on a transparent sheet, turn the sheet upside down and put it back on the table, you do get the mirror image. Can you explain why this is the case?

4.2 Combining knots

One way we may create a new knot from two given knots is to combine them in the manner of the two examples shown in figure 4.8.

(a)   (b)

Combined knots.

— Figure 4.8 —

Task **4.9** Construct the two knots in figure 4.8 on the previous page with
your rope.

Do you know the names of these knots and what they are used for?

Which knots have been combined to make each of the knots in figure 4.8?
Which of the resulting combined knots is amphichiral?

Composition

A knot which is formed by combining knots in this way is called the *composition, knot
sum* or *connected sum* of the original two knots.

To be precise, to compose two knots (see figure 4.9a):

1. remove a small arc from each knot, creating four new ends (figure 4.9b);
2. connect the ends in pairs (figure 4.9c) so that a single knot remains (figure 4.9d).

We choose the two removed arcs in such a way that they are not part of a crossing and the
new connections should clearly not create any new crossings.

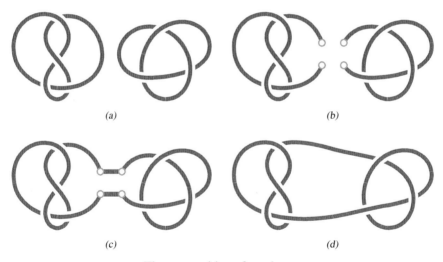

(a)   (b)

(c)   (d)

The composition of two knots

— Figure 4.9 —

Task **4.10** What happens if one of the two knots is the unknot? In this case,
does it matter where we remove the arcs?

In fact it does not matter where we remove the arcs, whatever two knots we are working
with. To see this, observe that you can always make one of the two knots very very tiny,

slide it to the new place where you want to add it and then make it big again. Hence it does not matter where you add the knot.

The symbol # is used for the composition of two knots, so that $K \# J$ denotes the composition of the knot K with the knot J. Here K and J are called the *factors* of the new knot.

As you found in task 4.10, composing a knot with the unknot leaves the knot unchanged. If the factors are non-trivial, that is, neither of them is the unknot, then the composition of two knots is called a *composite* knot.

> **Task 4.11**   How many ways are there to create a composite knot according to the procedure on the facing page?

There are some complicated knots which are not composite, and it is not always very easy to determine whether a complicated knot is composite or not.

> ✳ **Task 4.12**   Determine whether the knot in figure 4.10 is composite or not.

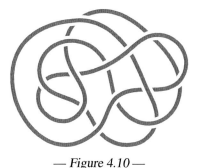

— Figure 4.10 —

Prime knots

Both of the knots shown in figure 4.8 on page 37 have crossing number 6. But there are also knots with crossing number 6 which are not composite, such as the one shown in figure 4.11 on the following page; indeed, there are exactly three such knots, as you will see on page 112 in the table at the back of the book.

A knot which is not the composition of two non-trivial knots is called a *prime* knot. It is a convention in knot theory that knot tables only list prime knots and this book follows this convention for the table on page 111. The name prime, of course, is used by analogy with the natural numbers, where a number that cannot be broken into smaller parts—the factors—is called a prime number or just a prime.

A knot with crossing number 6 that is not composite.

— *Figure 4.11* —

✳ **Task 4.13** Figure 4.12 shows two similar-looking knot diagrams with 10 crossings. However, one of the knots is prime and one is composite. Decide which is which, and determine the factors of the composite knot. [Hint: start by considering the knot in (a).]

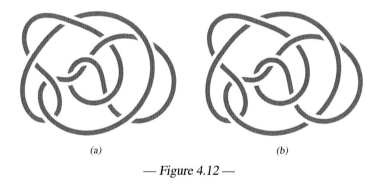

(a) (b)

— *Figure 4.12* —

When multiplying natural numbers, the number 1 plays a very special role: multiplying any other number by 1 has no effect. Mathematicians call 1 the *identity* for multiplication of natural numbers. How does this work for knots? Well, the unknot is also such an identity. If you take the composition of any knot with the unknot, the resulting knot is exactly the knot that you started out with.

But this still leaves us with the question: is the unknot a composite knot? It turns out that the answer is no: the unknot is not composite, it cannot be composed from two other knots. Unfortunately, we do not have the means to prove this here—to do so we would need to make a small excursion into the mathematical field of topology.

Task 4.14 Assuming that the unknot is not composite, explain why it is impossible to undo a given knot by composing it with another suitable knot.

Compare this to the situation for numbers: is it possible to multiply the number 5, say, by another natural number to obtain the identity 1?

Task 4.14 has a practical implication: if you have a hopelessly entangled rope which you are trying to disentangle, it will not help to add more knots.

There is another analogy between prime knots and prime numbers. When you multiply two primes, changing the order of the multiplication does not change the result (this is called the *commutative* property of multiplication): $5 \times 7 = 7 \times 5$. Exactly the same happens for knots: in the composition of two knots, the order is irrelevant. Whichever knot you write first, the resulting composite knot is the same: $K \# J = J \# K$. To see why, consider looking at the composition from behind: the right-hand knot is now on the left, and *vice versa*.

> **Task 4.15**   Which knots in figure 4.13 on the following page are prime and which are composite? When a knot is composite, determine its factors.

Composition and crossing numbers

For all the examples we have seen so far, composing a knot with crossing number n and a knot with crossing number m has yielded a new knot with crossing number $n + m$. This seems such an obvious fact that one does not really dare to question it. However, at the time of writing it is not known whether this is always the case, though the result is known to be true for knots with small crossing numbers, such as those listed in the table on page 111 and those considered in the next two tasks.

But the following general question remains unanswered: is it possible to create a composite knot whose crossing number is smaller than the sum of the crossing numbers of the factors, in other words, by composing two knots can we make crossings disappear?

> **Task 4.16**   There are exactly two composite knots with crossing number 7 (not counting mirror images). Sketch their knot diagrams and explain why there are no more than two.

> **Task 4.17**   There are quite a few ways to construct a composite knot with crossing number 10. Sketch a few such knot diagrams.
>
> When a composite knot with crossing number 10 is constructed, explain which crossing numbers the factors can have.

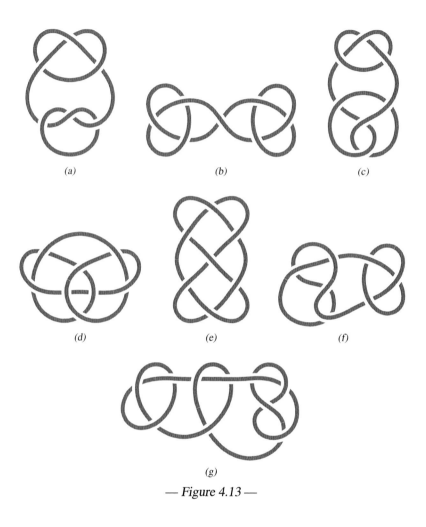

(a)   (b)   (c)

(d)   (e)   (f)

(g)

— *Figure 4.13* —

4.3   Changing crossings

We have already met the idea of changing the crossings in a knot diagram in section 4.1, where we observed that changing *all* the crossings produces a diagram of the mirror image of the knot. But suppose we only change some of the crossings, rather than all of them, what happens then?

Figure 4.14b on the next page shows the result of changing two crossings in the diagram of the trefoil in figure 4.14a. In this case we get a different knot, in fact the unknot. (We met these figures in passing on page 24, when discussing knot diagrams obtained from projections.)

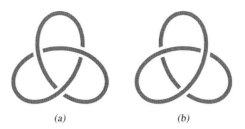

(a) (b)

Changing two crossings in the trefoil.

— *Figure 4.14* —

Task 4.18 What happens if only one crossing is changed in figure 4.14a?

We see that, in the case of the trefoil, changing crossings always results in either the unknot, or the mirror image. However, with some other knots it is possible to obtain a completely different knot by changing crossings.

Task 4.19 Figure 4.15a shows the figure of eight. Identify which crossings need to be changed to obtain the other two diagrams.

What are the resulting knots?

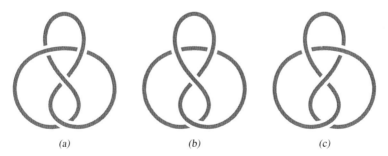

(a) (b) (c)

Changing crossings in the figure of eight.

— *Figure 4.15* —

The simple idea of changing crossings turns out to be a useful one, which re-occurs in various parts of knot theory. However, another related and at first sight appealing idea, turns out to have limited usefulness, as we see next.

The unknotting number

As we have seen, it may be possible to arrive at the unknot by changing some crossings in a given knot diagram.

> ❋ **Task 4.20**   Change the given number of crossings in each of the diagrams in figure 4.16 to produce a diagram of the unknot.

(a) Change two crossings.   *(b)* Change three crossings.

Change crossings to produce the unknot.

— *Figure 4.16* —

> **Task 4.21**   Using a pencil, draw a knot projection on a piece of paper. Now imagine laying down a piece of rope on the paper, starting where the pencil started and following the path of the pencil, eventually joining up the ends. Explain why the result is an unknot.
>
> Modify the crossings in the projection to obtain a knot diagram corresponding to the unknotted rope. Explain why this shows that it is always possible to change crossings in any knot diagram to produce a diagram of the unknot.

We may be tempted by the result of task 4.21 to define a new number associated with a knot, the smallest number of crossings that need to be changed in a diagram in order to unknot it. But each knot has many possible knot diagrams, and perhaps we can reach the unknot after changing fewer crossings in a different diagram. So to obtain a sensible definition, we have to consider all possible diagrams: we define the *unknotting number* of a knot to be the minimal number of crossings (across all diagrams) that need to be changed to get the unknot.

Notice that once we have found a diagram where only one crossing needs to be changed to get the unknot, then we know the uncrossing number is definitely 1—provided we are sure the knot itself is not the unknot—because no other diagram can need a smaller number of crossings to be changed. It was proved in 1985 that a knot with unknotting number 1 is a prime knot.

Task 4.18 on page 43 shows that the unknotting number of the trefoil is 1; from task 4.19 we see that the figure of eight also has unknotting number 1.

Task 4.22 Find the unknotting number of the knot shown in figure 4.17 (an *8-twist* knot).

— Figure 4.17 —

Task 4.23 Find the unknotting number of the cinquefoil and 3-twist knots (see figure 3.12 on page 27).

Unfortunately, the unknotting number can be extremely difficult to calculate. Your answer for the cinquefoil in task 4.23 is probably correct, but you cannot be certain—perhaps there is some diagram requiring fewer changes. Similarly, when doing task 4.20 you may very quickly have found which crossings to change; the hard part in finding the unknotting number is proving that no other diagram requires fewer changes. In fact the knots in figures 4.16a and 4.16b on the facing page were only confirmed to have unknotting numbers 2 and 3 in 1986 and 2005 respectively, using sophisticated techniques.

The unknotting number is still unknown for some knots with a relatively small crossing number: figure 4.18 shows an example with crossing number 10 for which, at the time of writing, all that is known is that the unknotting number is 2 or 3.

A knot with unknotting number 2 or 3.

— Figure 4.18 —

The unknotting number can also be counter-intuitive—there are knots for which, in order to get to the unknot with the fewest changes, you have to start from a diagram having more crossings than the crossing number, the minimum number possible.

As you may already have decided, the unknotting number is not a particularly useful property of a knot: not only is it hard to calculate, but many knots have the same unknotting number, so it does not help much in our quest to distinguish between knots. We need to look harder in order to find more helpful properties. In the next chapter we consider one possible approach—using colours.

INTERLUDE

The figure of eight

We have already met the figure of eight knot, in chapter 3. Two more knot diagrams of the figure of eight are shown below, a symmetrical 'square' version and a Celtic knot.

The figure of eight is also called Listing's knot after the nineteenth century mathematician Johann Benedict Listing, who was a student of Gauss and one of the earliest mathematicians to investigate knots. It was Listing who first coined the term topology for the branch of mathematics which includes knot theory. The study of knots was one of the earliest topological investigations.

The figure of eight is an important knot in climbing, sailing and many other activities. The knot has three great merits, being simple to tie, strong, and easy to untie even after being put under considerable strain. It is also a versatile knot with various practical forms.

In its basic form (shown above), the figure of eight is used as a *stopper knot* to prevent a rope slipping through a hole. This form of the knot is used in sailing.

When tied in a *bight*—a loop of rope—the figure of eight is used to form a fixed loop in the end of a rope (shown above). This form of the knot is used by climbers (for example, to attach a rope to a harness) and by anglers.

47

The figure of eight can also be used to create a *bend* (shown above) joining two ropes, which may be of different thicknesses. This form of the knot is sometimes called the Flemish bend. The bend is made by tying a figure of eight in one rope, then threading the second rope back through the knot, retracing the path taken by the first rope.

Considered mathematically, the figure of eight is the unique knot with four crossings, and thus is one of the simplest possible knots. As we saw in section 4.1 on page 33, unlike the simpler trefoil the figure of eight is equivalent to its mirror image, and is therefore the simplest amphichiral knot.

The figure of eight is amphichiral.

Two figures of eight may be 'interwoven' in several ways to form a more complicated knot. The decorative knot shown below is formed in this way. The result is a symmetrical twelve-crossing knot.

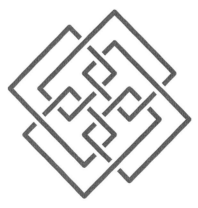

Two 'interwoven' figures of eight.

CHAPTER 5

Using colours

5.1 Knot invariants

In the previous chapter we learned the difference between prime knots and composite knots. In this chapter we will mainly study prime knots and will learn about another method of telling knots apart.

One method that can be used to distinguish mathematical objects is known as an *invariant*, which is a property that remains unchanged when transformations of a certain type are applied to the objects. In our case, we need to search for properties of knots which do not change when we deform the knot. We already know one such property, the (minimal) *crossing number*. We also saw in chapter 3 that the crossing number does not suffice to tell all knots apart, for example, for crossing number 6 we already know three seemingly different knots. There is clearly a need for other invariants.

How exactly can we define a knot invariant? We stated above that it is a property which does not change when one deforms the knot. But what does this mean? Of course, most of the time we want to consider the knot diagram instead of the knot itself. But deforming the knot is equivalent to deforming the knot diagram, and we know how to do this. There are only three ways to change a knot diagram, the three Reidemeister moves (see figure 2.12 on page 14), and every deformation of the knot corresponds to a series of Reidemeister moves. So in order to check that a property is a true knot invariant, we need to check that it remains unchanged under the three Reidemeister moves. This fact is so important in the remaining chapters of this book that we state it clearly once more:

Definition (Knot invariant)
If a property of a knot remains unchanged under the three Reidemeister moves, then the property is a knot invariant.

5.2 Three-colourability

We now define a new knot property and then check whether it really is a knot invariant. Consider a knot diagram to be divided into sections by the under-crossings. So each section is a piece of the "rope" between two under-crossings—a part of the diagram which can be drawn without a break. A knot is *three-colourable* (sometimes the term *tricolourable* is used) if each of its knot diagrams can be coloured according to the following rules:

- Each section is coloured in one of three different colours.

- At each crossing, either three different colours come together, or all the same colours come together.

- At least two colours are used.

Figure 5.1 shows two knot diagrams, each of which has been coloured with three colours (represented here by shades of grey). However, the colouring in figure 5.1b does not satisfy the above rules, since there are some crossings where only two colours appear (actually, in this case none of the crossings obey the rules).

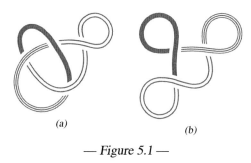

(a)

(b)

— Figure 5.1 —

Task 5.1 Explain why the colouring in figure 5.2 does not obey the above rules.

— Figure 5.2 —

Task 5.2   Copy figure 5.1 on the facing page and colour it with three colours, for example, red, blue and green. Why can you not make the knot diagram in figure 5.1b obey the rules by changing the colours in some way?

Task 5.3   Sketch a trefoil and a figure of eight and determine whether they are three-colourable or not. Is the unknot three-colourable?

In task 5.3 you should have found that the trefoil is three-colourable and the unknot is not three-colourable, but the figure of eight probably caused some problems, for good reason, since it is not three-colourable, as we shall see later in this chapter. We will also prove that three-colourability is a knot invariant. As a result we can be sure that the trefoil and the figure of eight are different knots. This is of course nothing new: we already showed in chapter 3 that they have different crossing numbers.

Before proving that three-colourability really is a knot invariant we will first consider how to use it to tell knots apart. Once we know that three-colourability is a knot invariant, if one of two knots is three-colourable and the other is not, then we know for certain that the two knots are not the same. One immediate consequence is that knots do exist! This follows from fact that the unknot is not three-colourable and the trefoil is not, so the trefoil is definitely knotted.

However, a note of caution is due. When two knots are both three-colourable, or both are not three-colourable, then we cannot deduce anything about the two knots. They could be different, but they could also be the same, three-colourability has nothing to say about the matter.

Task 5.4   For each pair of knot diagrams in figure 5.3, determine whether or not you can use three-colourability to distinguish between the two knots.

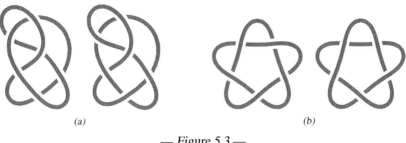

(a)   (b)

— *Figure 5.3* —

Three-colourability is a knot invariant

It is now time to prove that three-colourability is a true knot invariant. This means we have to check that the property of being three-colourable remains true if we change the knot diagram by making one of the Reidemeister moves.

Suppose we have a three-colourable knot diagram, and we change the diagram by applying a single Reidemeister move. You can imagine that this move takes place in a small part of the knot diagram, which we will encircle. We need to check that the rules of three-colourability are still valid after the move and that the ends of the rope that leave the circle stay the same colour (since outside the circle nothing has changed in the knot diagram). If this is guaranteed then we know that the knot diagram is still three-colourable after performing that particular Reidemeister move.

We will start with a 'twist' move of type I, shown in figure 5.4.

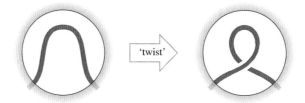

Three-colourability under a Reidemeister move of type I.

— Figure 5.4 —

Outside the circle in figure 5.4 the knot diagram remains unchanged. We assumed that the original knot diagram was three-colourable, so somewhere outside the circle at least one other colour appears, since this is not possible within the circle. When we now perform the Reidemeister move we create a new crossing. At this crossing our only option is to use the same colour three times, since the strands entering the circle from the right and the left have the same colour. But one colour at a crossing is permitted, so the knot diagram is still three-colourable after having performed the move. Hence we have proved that three-colourability is invariant under this type I Reidemeister move.

> **Task 5.5** Show that three-colourability is also invariant under the other possible type I Reidemeister moves (see figure 2.12a on page 14).

How about a Reidemeister move of type II? Consider the 'poke', where two strands are moved past each other and two new crossings are created. There are now two cases: either both strands have the same colour (figure 5.5a on the next page); or they have different colours (figure 5.5b).

When both strands are the same colour (figure 5.5a), we do not have to add a new colour after performing the move: the newly created strand can be the same colour as the four strands leaving the circle. Since we assumed that the diagram was three-colourable before

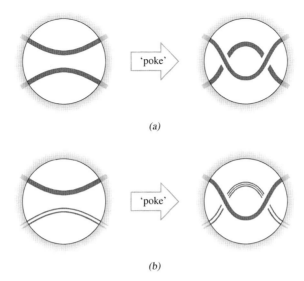

(a)

(b)

Three-colourability under a Reidemeister move of type II.

— *Figure 5.5* —

performing this move, once again at least one other colour appears somewhere outside the circle, and this is still the case after the move. Now the rules of three-colourability are still valid within the circle, so we have shown that three-colourability is invariant in this case.

In the second case (figure 5.5b) there are two strands of different colours at the start. After making the move there is only one way to colour the new strand without violating the rules. The four strands leaving the circle are of two different colours, so we already have two colours at the newly created crossings, therefore the new strand has to be given the third colour. This means that the rules for three-colourability are still valid after the move and therefore three-colourability is also invariant in this case.

We have dealt with both cases and therefore conclude that three-colourability is also invariant under this type II Reidemeister move.

Task 5.6   Show that three-colourability is also invariant under the 'unpoke', the other possible type II Reidemeister move (see figure 2.12b on page 14).

✻ **Task 5.7**   Prove that three-colourability is also invariant under a 'slide', a Reidemeister move of type III (see figure 2.12c on page 14). This is the hardest move to deal with, since there are six strands entering the circle and there are various ways to colour these strands; there are also four different versions of move to deal with. You will need to check all the possibilities.

If you succeeded with task 5.7, then that completes a proof that three-colourability is indeed a knot invariant.

Is a given knot three-colourable?

It is not always straightforward to ascertain whether a particular knot is three-colourable or not. If we are lucky, we can simply try to colour the knot; should we succeed we conclude that the knot is three-colourable. Should we fail, however, does this mean that the knot is not three-colourable? Not necessarily: perhaps we chose the wrong colour somewhere during the process, so we really need to try again. Of course, we cannot just keep trying, we need to find a systematic approach, as illustrated by the following example.

Is the rather complicated knot shown in figure 5.6 three-colourable?

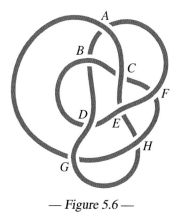

— *Figure 5.6* —

We have labelled each crossing with a letter simply to be able to explain our approach, this is not a necessary part of the method. We did this randomly; the order is not important, we just want to be able to refer to a particular crossing or strand. If you really want to understand our explanation, make a few copies of the figure, including the letters, and have three coloured pens ready. It will be easiest if you use the same colours as we do—red, green and blue—but any other colours are allowed, of course.

1. Choose a starting crossing, A say, and colour the three strands at this crossing with three different colours. For example, AB red, $GACE$ blue and $AFED$ green.

2. Is there any crossing where the third colour is forced by the two other strands? In our case this happens at crossing E: the strand EH has to be red.

3. Now choose the next crossing, B say. Colour CBD green, so $BDGH$ has to be blue.

We now see that there are only two colours at crossing D, namely blue and green. This means our colouring is not correct. However, we made a choice at step 3, where we chose to make CBD green. Perhaps this choice was wrong, so let us try something else:

3. Colour CBD blue; then $BDGH$ has to be green.

It is probably clear to you that interchanging blue and green does not solve the problem at crossing D. However, we do have one other option at step 3:

3. Colour CBD red; then $BDGH$ has to be red too.

Once again things go wrong at crossing D, but this time two red and one green strand meet. Does this mean the knot is not three-colourable? No, we cannot be certain yet. Right at the beginning we made a choice, namely, to use different colours for the three strands at crossing A. What happens if we make the alternative choice?

1. Choose only one colour at crossing A, you can, for example, colour AB, $AFED$ and $GACE$ red.

2. Now we are forced to colour EH red too.

3. Choose a crossing where you would like to continue, B say. Colour CBD blue and $BDGH$ green (we want to avoid simply colouring the whole diagram red).

4. Now we have no choice left at crossing G, where GHF has to be blue.

5. And then we have no choice left at crossing F, where CF has to be green.

— *Figure 5.7* —

Now let us check the whole knot (figure 5.7). All strands are coloured, we have used at least two different colours (we have actually used three), and at each of the crossings either one or three colours meet. This means that this knot is, after all, three-colourable!

The moral is, do not judge the situation too quickly. You can only say that a knot is *not* three-colourable if you have tried all possibilities. And in order to make sure you have not forgotten to consider some possibility you need to use a systematic approach.

Task 5.8   Show that the figure of eight knot is not three-colourable.

Task 5.9   Use a three-colourability argument to show that the trefoil and the figure of eight are different knots.

Task 5.10   Show that all the knots in figure 5.8 are three-colourable.

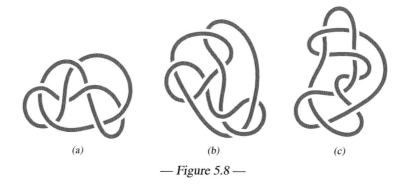

(a)   (b)   (c)

— *Figure 5.8* —

Task 5.11   Consider the three knots with crossing number 6 on page 112 in the table at the back of the book. Does three-colourability allow you to be certain whether one of the three is different from the other two?

Task 5.12   Explain why it is not possible to use three-colourability to tell three given knots apart.

✳ **Task 5.13**   Determine which of the knots in figure 5.9 on the next page are three-colourable and which are not.

In this chapter we have learned what mathematicians mean by a knot invariant and have become familiar with one such invariant, three-colourability. The idea was introduced around 1960 and has been generalised to allow the use of more than three colours, so it is possible to talk about a knot being, say, five-colourable.

Three-colourability can often be used to determine whether or not two knots are different, but it does not suffice to tell all knots apart. Even a crossing number as low as 6 causes problems: in task 5.11 we saw that one of the three knots with crossing number 6 is three-colourable and the other two are not. So three-colourability shows that one is definitely different to the other two, but does not distinguish between these two. How can we distinguish them? We clearly need better and stronger knot invariants.

However, there is one important consequence of our work on three-colourability. We have shown that the trefoil is three-colourable and that the unknot is not. It follows that the

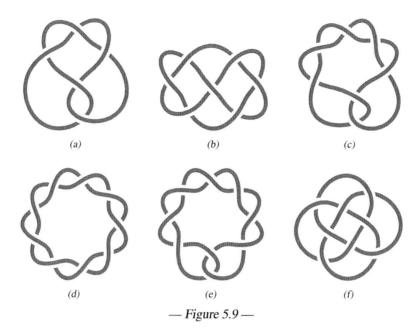

(a)   (b)   (c)

(d)   (e)   (f)

— Figure 5.9 —

trefoil and the unknot are definitely different, so there is no way to undo a trefoil by moving the strands around. We took this for granted in chapter 3, relying on the fact that the trefoil really did have crossing number three, but we never actually proved that.

In chapter 7 we will investigate another knot invariant, which will allow us to distinguish many more knots—but not all! First, in chapter 6, we describe an important generalisation of the idea of a knot.

Hunter's bend

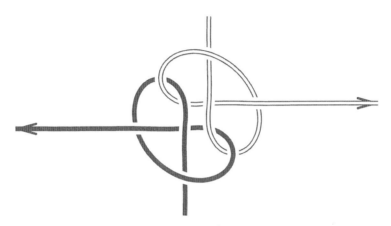

Mankind has used knots for thousands of years, so finding a 'new' one is remarkable. Hunter's bend, first described in the twentieth century, featured on the front page of *The Times* on 6 October 1978. According to the article this was a newly invented knot, credited to Dr Edward Hunter who had used it for years to tie broken shoelaces. The resulting publicity led to the formation of the International Guild of Knot Tyers. It later emerged, however, that the knot had already been described in *Knots for Mountaineering* [10] by Phil Smith, who called it simply a *rigger's bend*.

Hunter's bend has the form of two interwoven overhand knots. There is a whole family of such knots, ranging from the fisherman's knot to the alpine butterfly bend.

Fisherman's knot   Alpine butterfly bend

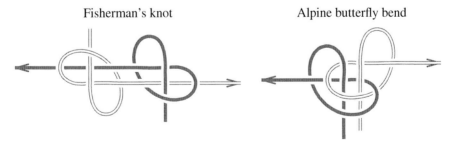

Bends are used to join two ropes; Hunter's bend is particularly useful for slippery ones, notably man-made fibres like polypropylene or nylon. The knot is symmetrical, so that the ends may be left long for supplementary use, such as the attachment of other lines, but since the knot is quite bulky it is not really suitable when the join has to pass through an eye such as a guide or pulley.

How to tie Hunter's bend

Lay the ropes end to end.

Pick up a double loop, by lifting and twisting.

Thread one end through the loop from the opposite side.

Repeat with the other end; carefully work the knot tight.

CHAPTER 6

Links

6.1 What is a link?

In this chapter we describe an important generalisation of the idea of a knot. Consider the diagram in figure 6.1a. At first sight this appears to represent a knot, but a more careful look reveals something else. As indicated in figure 6.1b, this "knot" is actually made from two pieces of rope.

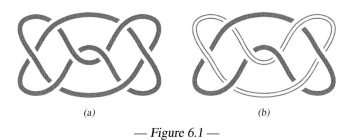

(a)   (b)

— Figure 6.1 —

So far we have always considered just one knotted piece of rope. But perhaps we should also consider two, or even more, interlinked pieces of rope? Indeed, there is at least one very good reason to do so: one common use of a "knot" in real life is to join two ropes. The purpose of this chapter is to extend our ideas about knots to more than one piece of rope.

However, we do need to be precise about what we are going to consider, just as we were for knots in section 2.2 on page 8. Once again, for example, we want to exclude loose ends. The simplest way to be precise is to require that each piece of rope, considered separately, itself forms a knot (which may be the unknot). In mathematics we call several knots tangled together a *link*. For any link we can draw a link diagram, in the same way that knot diagrams are drawn (see section 2.3 on page 9).

Task 6.1 Determine which diagrams in figure 6.2 represent a knot and which represent a link.

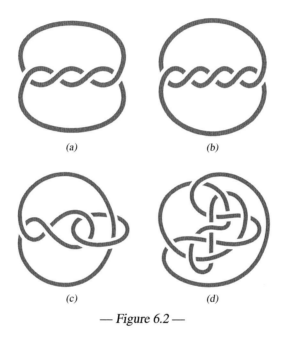

(a)

(b)

(c)

(d)

— Figure 6.2 —

Much of what we have said so far about knots can also be said about links. For example, we call two links *equivalent* if one link diagram can be deformed into the other by a sequence of Reidemeister moves.

For example, one of the simplest links is the *Whitehead link* (see figure 6.3 on the facing page), first studied in 1934 by the British mathematician J. H. C. Whitehead[*]. One of the links we have just considered, that in figure 6.2c, is equivalent to the Whitehead link.

❋ Task 6.2 Show with the help of Reidemeister moves that the two diagrams in figures 6.2c and 6.3 represent the same link.

We may define the concepts of *composition* of links, of a *prime* link and of a *composite* link, in the same way that we did for knots (see section 4.2 on page 37). The table on page 111 includes all prime links with small crossing numbers.

---

[*]Whitehead was working on the Poincaré conjecture, a question in topology posed in 1904 by the French mathematician Henri Poincaré. The question turned out to be extraordinarily difficult, but nearly a century later Grigori Perelman presented a proof in three papers in 2002 and 2003.

The Whitehead link.

— *Figure 6.3* —

As it happens, every diagram in figure 6.2 on the preceding page which represents a link uses only two pieces of rope. It is, of course, possible to use more than two and figure 6.4 shows a well-known example, the so-called *Borromean rings*, a link consisting of three rings.

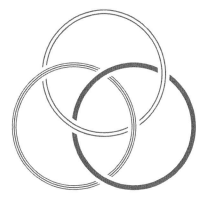

The Borromean rings.

— *Figure 6.4* —

The link gains its name from the Borromeo dynasty of the Italian Renaissance, who used it in their family crest [14]. History apart, the link has an interesting mathematical property: the three rings are clearly linked together—they cannot be separated—however, removing any one of them unlinks the other two as well. Perhaps this was significant to the Borromeos, implying that the family could only be strong by standing together?

The link shown in figure 6.5 on the following page has the same property—if you open one of the rings, the other two also become separated. In fact this link is the same as the Borromean rings (see task 6.3).

> ✳ **Task 6.3**   Show that the link in figure 6.5 is equivalent to the Borromean rings, by finding the Reidemeister moves required to go from a diagram of the link (figure 6.6) to the 'standard' form given in figure 6.4.

— Figure 6.5 —

A diagram for the link in figure 6.5.

— Figure 6.6 —

6.2 Components

Each separate piece of rope used to form a link is called a *component* of the link. A link using two pieces of rope, like those in figures 6.2a, 6.2c and 6.2d on page 62, is called a *2-component link*, or a *link with two components*. The Borromean rings form a link with three components. And a knot is nothing other than a link with one component.

Task 6.4   How many components has each link in figure 6.7?

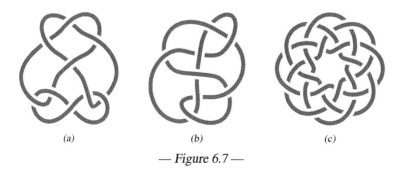

(a)   (b)   (c)

— Figure 6.7 —

The approach we take in this book—study knots and consider links as a generalisation—is the traditional one. However, it is perhaps more natural to study links as a whole, and to consider a knot as a special type of link. Recently this approach has been very useful in proving results about knots.

There is enough material to fill a whole book just about links. Such a book would discuss the various methods of distinguishing links, and would thus include definitions of new invariants for links. We do not intend to discuss this topic in detail—we do not want to write a book about links—but there are some concepts from the theory of links that are helpful when studying knots. We describe one of these next.

6.3 The linking number

The simplest invariant of links is the obvious one, the number of components. And it is clear that two links with different numbers of components cannot be deformed into each other. Hence the links in figure 6.7 on the preceding page are definitely all different, because they have different numbers of components. However, that is not the end of the story. Consider the two links in figure 6.8, each of which is a 2-component link. Figure 6.8a shows the *unlink with two components*, figure 6.8b shows the *Hopf link*.

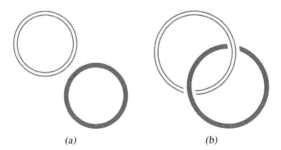

(a)   (b)

The unlink and the Hopf link.

— Figure 6.8 —

In this case it seems clear, at least intuitively, that these links cannot be equivalent, even though they have the same number of components—you cannot separate the Hopf link, whereas the unlink is obviously separated. For more complicated links, however, it is not so easy to determine how *linked* they are. What we need is a method to measure the degree of 'linked-ness'. To achieve this we define the *linking number*. Of course, it is not enough just to make a definition, we must also show that we really do have a proper invariant of links. We therefore need to show that the linking number remains unchanged under each of the three Reidemeister moves. (As we stated on page 62, two links are equivalent if one link diagram can be deformed into the other by a sequence of Reidemeister moves.)

We shall explain how to calculate the linking number by considering a link with two components (see figure 6.9 on the next page), but the method applies equally well to a link with any number of components. Firstly we choose an *orientation* for the link, which means that we choose a direction for each component. We indicate each of these directions by arrows, as shown in the diagram.

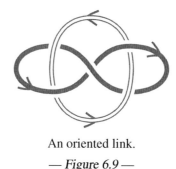

An oriented link.

— Figure 6.9 —

Once we have chosen an orientation, we label every place where two components cross either +1 or −1. There are only two possible types of crossing, those shown in figure 6.10 (note that the actual colours of the strands are irrelevant). A crossing like that in figure 6.10a is labelled +1 and one like that in figure 6.10b is labelled −1, as shown.

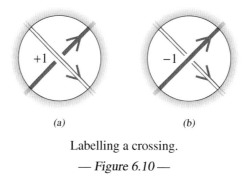

(a)   (b)

Labelling a crossing.

— Figure 6.10 —

Task 6.5   Try—before you continue reading—to formulate in words how to distinguish the two types of crossing shown in figure 6.10.

There are various ways to determine the nature of a crossing. One way is to imagine you are travelling in a train on the 'overpass' in the direction of the arrow. As you cross the bridge, if you see another train coming from the right, then use the label +1; if the other train is coming from the left, then use the label −1.

Alternatively, the types of crossing can be described in terms of rotations (see figure 6.11 on the facing page). For the crossing in figure 6.11a, the strand on top has to be rotated in an anticlockwise sense so that the strand and the orientation both match the strand below. In mathematics, anticlockwise is considered to be positive so the crossing is labelled +1. In figure 6.11b, the top strand has to be rotated clockwise, which is considered to be negative, so the label is -1. Note that we always use the smallest possible rotation.

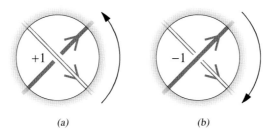

(a)   (b)

Finding the label of a crossing from rotations.

— Figure 6.11 —

Remember that we only label crossings where the strands belong to different components of the link—crossings where both strands belong to the same component of the link are left unlabelled.

We can now define the *linking number* of a link. To find the linking number, add all the labels, that is, all the $+1$s and -1s, and divide the result by two.

> **Task 6.6**   From its appearance, explain why the link in figure 6.9 on the preceding page may be considered to be linked twice. Show that this link has linking number 2
>
> Hence explain why the definition of the linking number includes the phrase 'divide by 2'.

> **Task 6.7**   What is the linking number of a knot?

> **Task 6.8**   Choose an orientation for the link in figure 6.12 on the following page and calculate its linking number.
>
> Now reverse the orientation of just one of the components. What happens to the linking number?
>
> What happens if you change the orientation of both components?

As you should have found from task 6.8, it is possible for the linking number to change when the orientation of one component is reversed. So the linking number is definitely *not* an invariant of links. But this still leaves open the question: is the linking number an invariant for *oriented* links? Now because two links are equivalent if one link diagram can be deformed into the other by a sequence of Reidemeister moves, in order to answer

— Figure 6.12 —

this question we need to consider the three Reidemeister moves (see figure 2.12 on page 14).

For a Reidemeister move of type I the answer is immediate since the linking number only counts crossings between two different components, whereas a move of type I involves only one of the components. So the linking number does not change under a move of type I.

For a Reidemeister move of type II a little more thought is required. Clearly, if two strands from the same component are poked or unpoked, then nothing changes. But it is possible for strands from two different components to be involved. So suppose we have two strands belonging to different components, and let us choose an orientation.

> **Task 6.9**   Describe—before you continue reading—what will happen to the linking number if you perform a Reidemeister move of type II.
>
> What would happen if one of the components had been given the reverse orientation?

The linking number does not change under a move of type II, even though either two new crossings come into existence or two crossings disappear. This is because one of the two crossings is always labelled +1 and the other −1, no matter how the orientations are chosen. Figure 6.13 shows one case.

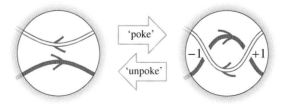

The linking number under a Reidemeister move of type II.

— Figure 6.13 —

Task 6.10   Show that the linking number is invariant under a Reidemeister move of type III.

If you managed to solve task 6.10 successfully, then that completes a proof that the linking number is an invariant of oriented links.

Normally, however, links do not come with a natural orientation. Often, therefore, only the absolute value of the linking number is considered, in other words the sign is omitted. When modified in this way the linking number becomes an invariant for links with two components (without orientation).

Task 6.11   Use the linking number to prove that the 2-component unlink and the Hopf link (figure 6.14) are not equivalent.

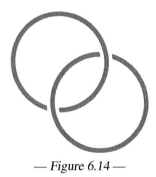

— *Figure 6.14* —

Task 6.12   Use the absolute value of the linking number to show that the two links in figure 6.15 are different.

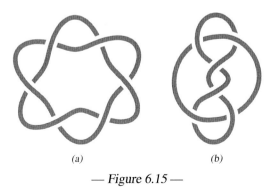

(a)   (b)

— *Figure 6.15* —

Task 6.13 Calculate the linking number of the Whitehead link (figure 6.16) and compare this to the linking number of the unlink with two components.

— Figure 6.16 —

Task 6.14 The famous German Renaissance painter Albrecht Dürer (1472– 1528) created a series of engravings of decorative knots. Figure 6.17 shows a link used by Dürer as part of *Der zweite Knoten*. What is the linking number of this link?

A link used by Dürer.

— Figure 6.17 —

6.4 Three-colourability

When we looked at knots we saw that the crossing number by itself is not enough to tell all knots apart. A similar situation occurs for links: even taken together, the number of components, the crossing number and the linking number do not distinguish all links. For example, each of the links in figure 6.12 on page 68 and figure 6.17 has two components, crossing number 9 and linking number 3, but they are different links. In order to distinguish links in general, mathematicians need to make use of other invariants.

Perhaps surprisingly, the knot invariant three-colourability (defined in section 5.2 on page 50) is also an invariant for links.

Task 6.15 Show that the unlink with two components is three-colourable (in contrast to the unknot).

Show that the unlink with *n* components, where *n* > 2, is three-colourable.

Task 6.16 Show that the Whitehead link (figure 6.16 on page 70) is not three-colourable and is therefore not equivalent to the unlink with two components.

Task 6.16 once again shows the strength of three-colourability. Although neither the number of components nor the linking number can distinguish the Whitehead link and the unlink, three-colourability is able to do so!

Task 6.17 Show that the link in figure 6.18 has two components, but is not equivalent to the Whitehead link.

— Figure 6.18 —

✳ **Task 6.18** Show that the links in figure 6.19 are different.

— Figure 6.19 —

Torus knots

Suppose you wind a rope around a glass ring, then join the ends of the rope, as shown below.

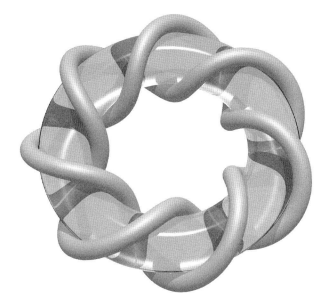

Now smash the glass, leaving an entangled rope. (This is a thought experiment, we do *not* advise you to use an actual glass ring and smash it!) The result is a *torus knot*.

In the figure above, the rope passes through the hole seven times whilst going twice around the ring. When the rope passes through the hole p times whilst going around the ring q times, the result is called the (p, q)-torus knot. So the figure shows how to make a $(7, 2)$-torus knot.

Torus knots are often very attractive. Indeed, a torus knot underlies the process used to make some decorative knots such as Turk's Heads (see [7], for example).

Some torus knots are already very familiar.

Activity 1   Use a piece of rope to make a $(3, 2)$-torus knot. Do you recognise the resulting knot?

Making a torus knot can sometimes lead to unexpected results.

Activity 2   Use a piece of rope to make a $(3, 1)$-torus knot. What is the resulting knot?

Sometimes two torus knots obtained in different ways are actually the same.

Activity 3   Use a piece of rope to make a $(2, 3)$-torus knot. Do you recognise the resulting knot?

The $(3, 2)$- and $(2, 3)$-torus knots.

As it happens, our definition is not quite right.

Activity 4   Try to make a $(4, 2)$-torus knot. What do you notice?

Activity 4 shows that sometimes more than one rope is required and a link is formed rather than a knot. Even so, the term 'torus knot' is still used in this case.

Several of the knots and links that appear elsewhere in this book are also torus knots. For example, the cinquefoil knot (see figure 3.12 on page 26) is the $(5, 2)$-torus knot, and the Hopf link (see figure 6.14 on page 69) is the $(2, 2)$-torus knot.

The $(4, 3)$-, $(5, 3)$- and $(6, 2)$-torus knots.

CHAPTER 7

Knot polynomials

We have now met various invariants of knots and links: the crossing number, three-colourability, the unknotting number and the linking number. None of them is ideal, since there are some knots which they cannot tell apart. In particular, we have not yet been able to distinguish the trefoil and its mirror image, something we promised to do in section 4.1 on page 33. Nor have we been able to fully classify knots with crossing number 6 (see task 5.11 on page 56).

We clearly need a more powerful invariant, and this chapter discusses one of the most successful and most interesting methods of distinguishing two knots—*knot polynomials*. This requires quite a bit of work, and at times you may find that this chapter is harder than earlier ones.

As the name indicates, a knot polynomial is an invariant which associates a polynomial to every knot. (We actually generalise the concept of polynomials to include negative powers, so that an expression like $x^7 - x^3 - x^{-5}$ is possible.) The key property of a knot polynomial is the following.

Property: *If two knots have different polynomials, then we can be certain that the knots are different too.*

Our aim in this chapter is to describe the so-called *Jones polynomial*, discovered in 1984 by the New Zealand mathematician Vaughan Jones (figure 7.1). In 1990 Jones was awarded the Fields medal* for this discovery at the international mathematics conference in Kyoto, Japan.

In fact the Jones polynomial was the second knot polynomial to be discovered. The first, the *Alexander polynomial*, was discovered in 1923 by the American mathematician James Alexander. We will not describe the Alexander polynomial here since the Jones polynomial

---

*The Fields medal is the highest award given in mathematics, comparable to the Nobel prize in the sciences.

Vaughan Jones in May 2011.

— Figure 7.1 —

is more powerful. For example, the Alexander polynomial never distinguishes a knot and its mirror image, whereas in most cases the Jones polynomial does.

The way we present the Jones polynomial in this chapter is very different to Jones' original approach, which is beyond the scope of this book. Before we go on to describe the Jones polynomial, it will be helpful to study a simpler polynomial. Unfortunately, as we shall see, this one is not a true knot invariant.

7.1   The bracket polynomial

In this section we will show how to derive a suitable polynomial from first principles. This will not be the Jones polynomial, but a simpler one.

What properties should the polynomial have?

- ✑ We want to associate a polynomial to every knot (or even to every link).

- ✑ We wish to determine the polynomial from the knot diagram.

- ✑ In order for the polynomial to be a knot invariant, we should get the same result if we start with a different knot diagram of the same knot, that is, the polynomial should not change under the Reidemeister moves (see figure 2.12 on page 14).

If we succeed in constructing such a polynomial, then we can be sure that two knots with different polynomials are different knots.

Suppose we denote a given knot by K. We shall denote the polynomial of the knot K by $\langle K \rangle$ and will refer to it as the *bracket polynomial* of K. The polynomial is also known as the *Kauffman bracket* because it was discovered by Louis Kauffman in 1987 at the University of Chicago.

It seems sensible to start with the simplest case and ask, what polynomial should we associate to the unknot? The simplest possible polynomial we can select is a constant polynomial—one which always has the same value—for example, the polynomial 1. (We could also have chosen 0, but we will see later that this would not make much sense). This gives us our first rule.

Rule 1: $\langle \bigcirc \rangle \equiv 1$.

Here the symbol $\bigcirc$ stands for the unknot. We use the symbol '$\equiv$' (equivalent) rather than '=' (equals) to indicate that we are dealing with polynomials; think of '$\equiv$' as meaning 'equal for all values of the variables'. So rule 1 says that the bracket polynomial of the unknot is just the polynomial with constant value 1.

What about other knots? We cannot just arbitrarily assign a polynomial to each and every knot since we want the polynomial to be determined by the knot. Indeed, to be practical, we need a method of finding the polynomial from a knot diagram. And because a given knot may have more than one diagram, we shall therefore need each polynomial to remain the same after a Reidemeister move.

What we would really like to have is a method of finding the polynomial of a complicated knot in terms of the polynomials of simpler knots—that would allow us to build up the polynomials in a systematic way. One possible approach is to consider what happens when we simplify a knot by removing a crossing. If we can determine how the associated polynomial is affected then we can proceed as follows: change one crossing after another into a non-crossing until the complicated knot is reduced to a collection of unknots that are not linked together—an unlink. Then we will be able to determine the polynomial of the original knot once we know the polynomial of an unlink. Since this process is likely to introduce extra components, we are certainly going to need to work with links, rather than just knots, in what follows.

The question remains, how should we eliminate crossings in this process? As shown in figure 7.2 on the next page, we can remove a crossing in two different ways. In each case the crossing is replaced by two uncrossed strands, whilst leaving the remainder of the knot unaltered. The two cases are determined by which pairs of strands are joined together (since there are only three ways to divide four points into two pairs, one of which corresponds to a crossing, it is clear that we have dealt with all possibilities).

Each time we do this we create two new knots, possibly links, with simpler diagrams than the original, in that each of them has one fewer crossing. But what effect should this have on the polynomial of the knot? Kauffman's ingenious idea was to combine the polynomials of these two simpler knot diagrams according to the following rule.

Rule 2: $\left\langle \times \right\rangle \equiv x \left\langle \,)(\, \right\rangle + y \left\langle \asymp \right\rangle$.

Two ways to remove a crossing.

— Figure 7.2 —

The right-hand side of Rule 2 has the form "x times one polynomial plus y times the other polynomial". Here x and y are two variables, which for now may be considered to be independent of each other (though we will see later that they are in fact related). The dotted circles indicate the region of the knot that is changed; the rest of the knot lies outside the circle, but remains unchanged when the crossing is removed.

For example, suppose the ends of the crossing are just connected together at the top and at the bottom, without any further crossings. Applying the rule to this simple knot diagram we get

$$\left\langle \infty \right\rangle \equiv x \left\langle \infty \right\rangle + y \left\langle \ominus \right\rangle.$$

Now the polynomial we obtain should not depend on how we look at the knot diagram. In particular, if we look at the crossing on the left of rule 2 with our head on one side we should get the same result. We therefore have the following variation of rule 2, obtained by rotating each of the three figures through 90°.

Rule 2a: $\left\langle \times \right\rangle \equiv x \left\langle \asymp \right\rangle + y \left\langle)(\right\rangle.$

Rules 2 and 2a allow us to remove the crossings one at a time. Each time we apply a rule we double the number of knots or links which have to be considered, though they have fewer crossings, and we get more complicated algebraic expressions. Eventually we obtain a collection of links, each of which consists only of unlinked loops. In other words, we have reduced the knot to a collection of unlinks. The following rule allows us to find the polynomial associated to an unlink.

Rule 3: $\langle L \sqcup \bigcirc \rangle \equiv z \langle L \rangle.$

The new symbol $\sqcup$ denotes the so-called *split union*. The split union of two knots or links K and L is obtained just by taking their union, but ensuring that the two knots are not linked in any way. We can do this by ensuring that one knot is placed inside some sphere, whilst the other is outside the sphere.

> **Task 7.1**   Use rules 1 and 3 to find the bracket polynomial of the unlink with two components.

When forming a split union, it clearly does not matter which knot is placed inside the sphere and which outside, so we also have the following rule.

Commutative law: $\langle K \sqcup L \rangle \equiv \langle L \sqcup K \rangle$.

With these rules, we can now simplify the knot gradually, at the same time forming a polynomial in the three variables x, y and z (we never introduce more variables). In the end we should obtain an expression where each bracket contains the unknot—whose bracket polynomial we know—so we can replace the final brackets by the number 1 and we will have found the bracket polynomial of our knot K.

In practice, we don't need to reduce to unknots. Since we know the polynomial of an unlink, using results like that of task 7.1, as we go along we can replace brackets containing unlinks directly by the corresponding polynomial.

Summarising, we have the following procedure.

To calculate the bracket polynomial: *Start with the given knot and use the rules to eliminate the crossings one by one. Each time you apply a rule the polynomial becomes more complicated and the knot simpler, until you are left with an unknot or an unlink. It is not hard to determine the bracket polynomial of either of these.*

Task 7.2   Determine the bracket polynomial $\left\langle \vcenter{\hbox{\includegraphics{figure8}}} \right\rangle$.

At the moment our bracket polynomials have three variables. Our goal is to ensure that the polynomial is invariant under as many Reidemeister moves as possible—three would be ideal! As we shall see, to achieve this it is necessary for the three variables to be related. Let us consider Reidemeister moves of type II. If the bracket polynomial is to be invariant under a move of this type, then we should have

$$\left\langle \vcenter{\hbox{\includegraphics{rm2a}}} \right\rangle \equiv \left\langle \vcenter{\hbox{\includegraphics{rm2b}}} \right\rangle. \tag{$*$}$$

Here is our first opportunity to do some calculations. Let us start with the left-hand bracket above and see what happens when we apply the rules (see calculation A). Since this is a completely new way of doing algebra we provide a step-by-step commentary.

Calculation A

Is the bracket polynomial invariant under a Reidemeister move of type II?

$$\left\langle \vcenter{\hbox{\includegraphics{a}}} \right\rangle \equiv x \left\langle \vcenter{\hbox{\includegraphics{b}}} \right\rangle + y \left\langle \vcenter{\hbox{\includegraphics{c}}} \right\rangle \tag{A.1}$$

$$\equiv x \left[x \left\langle \vcenter{\hbox{\includegraphics{d}}} \right\rangle + y \left\langle \vcenter{\hbox{\includegraphics{e}}} \right\rangle \right] + y \left[x \left\langle \vcenter{\hbox{\includegraphics{f}}} \right\rangle + y \left\langle \vcenter{\hbox{\includegraphics{g}}} \right\rangle \right] \tag{A.2}$$

$$\equiv x^2 \left\langle \underset{\smile}{\frown} \right\rangle + xyz \left\langle \underset{\smile}{\frown} \right\rangle + xy \left\langle \,\underset{\frown}{)\,(}\, \right\rangle + y^2 \left\langle \underset{\smile}{\frown} \right\rangle \qquad (\text{A.3})$$

$$\equiv \left[x^2 + xyz + y^2 \right] \left\langle \underset{\smile}{\frown} \right\rangle + xy \left\langle \,)\,(\, \right\rangle. \qquad (\text{A.4})$$

Commentary

In line A.1, rule 2a has been applied to the top crossing, thus eliminating one crossing but leaving the other crossing unchanged. So a crossing has been removed, at the cost of introducing the variables x and y.

In line A.2, the bottom crossing has been eliminated using rule 2 twice, once for each of the brackets in line A.1. On each occasion the bracket has been replaced by an expression with two brackets, so there are now four bracket terms altogether. If you were to expand everything, then there would be terms in x^2, y^2 and xy, so the whole expression is quadratic.

In line A.3, the unknot has been removed from the second bracket using rule 3, thereby introducing the variable z. The whole expression has also been simplified, removing the square parentheses, by multiplying out.

Finally, in line A.4, the bracket terms have been collected together.

Now the result of calculation A is not what we expected. For the bracket polynomial to be invariant under a Reidemeister move of type II, comparing with identity ($*$) on the preceding page, we see that line A.4 should consist of just the last bracket and nothing else. We can achieve this by setting the coefficient of the first bracket in line A.4 to be 0 and the coefficient of the last bracket to be 1. We therefore obtain the following pair of equations:

$$x^2 + xyz + y^2 = 0$$
$$xy = 1.$$

Solving the second equation for y gives

$$y = x^{-1}.$$

Substituting this expression for y in the first equation, we obtain

$$x^2 + xx^{-1}z + x^{-2} = 0,$$

which gives

$$z = -x^2 - x^{-2}. \qquad (**)$$

We have therefore expressed y and z in terms of x, which means that the polynomial does not actually need to use three variables, but can use only one. We shall use x, but could just as well have chosen y or z. Note the appearance of negative powers, which explains why we need to generalise the concept of polynomials to include these.

> **Task 7.3** Use the result of task 7.1 and equation (✱✱) to write down the bracket polynomial of the unlink with two components as an expression in x alone.

Let us reformulate our rules using only the single variable x. As before, L indicates an arbitrary knot or link.

Rule 1: $\langle \bigcirc \rangle \equiv 1$.

Rule 2: $\left\langle \vcenter{\hbox{$\times$}} \right\rangle \equiv x \left\langle \vcenter{\hbox{$)($}} \right\rangle + x^{-1} \left\langle \vcenter{\hbox{$\asymp$}} \right\rangle$.

Rule 2a: $\left\langle \vcenter{\hbox{$\times$}} \right\rangle \equiv x \left\langle \vcenter{\hbox{$\asymp$}} \right\rangle + x^{-1} \left\langle \vcenter{\hbox{$)($}} \right\rangle$.

Rule 3: $\langle L \sqcup \bigcirc \rangle \equiv [-x^2 - x^{-2}] \langle L \rangle$.

We also have the commutative law, of course.

Rules 2 and 2a are easily confused, so you may prefer the following verbal description that encapsulates both of them.

Rule 2/2a: *Replace a crossing by the two possible pairs of uncrossed strands. Suppose you are approaching the original crossing on the 'overpass'. Include a coefficient x when the replacement strand turns to the left, and a coefficient x^{-1} when it turns right.*

As the next task shows, our efforts now start to bear fruit: the bracket polynomial is indeed invariant under a Reidemeister move of type III.

> **Task 7.4** Show that the bracket polynomial is invariant under a Reidemeister move of type III.

We have now seen that the bracket polynomial is invariant under a Reidemeister move of type II or III. Before turning our attention to a Reidemeister move of type I, it is useful to calculate the bracket polynomial for a few more complicated knots and links, starting with the unlink with more than 2 components.

> **Task 7.5** Find the bracket polynomial of the unlink with 3 components.
>
> Now do the same for the unlink with n components.

Let us find the bracket polynomial of a more complicated link, namely the Hopf link shown in figure 7.3 on the next page. See calculation B—once again it will repay you to study every step carefully.

The Hopf link.

— Figure 7.3 —

Calculation B

What is the bracket polynomial of the Hopf link?

$$\left\langle \text{⬚} \right\rangle \equiv x \left\langle \text{⬚} \right\rangle + x^{-1} \left\langle \text{⬚} \right\rangle$$

$$\equiv x \left[x \left\langle \text{⬚} \right\rangle + x^{-1} \left\langle \text{⬚} \right\rangle \right] + x^{-1} \left[x \left\langle \text{⬚} \right\rangle + x^{-1} \left\langle \text{⬚} \right\rangle \right]$$

$$\equiv x \left[x \left(-x^2 - x^{-2} \right) + x^{-1} \right] + x^{-1} \left[x + x^{-1} \left(-x^2 - x^{-2} \right) \right]$$

$$\equiv -x^4 - 1 + 1 + 1 - 1 - x^{-4}$$

$$\equiv -x^4 - x^{-4}.$$

Task 7.6   For the first three lines of calculation B, write down which rule or rules have been used.

Notice that all three rules are used when calculating the bracket polynomial for the Hopf link.

Task 7.7   Consider why the mirror image of the Hopf link has exactly the same bracket polynomial as the Hopf link, even though different rules are used in the calculation.

If we knew that the bracket polynomial was a true knot invariant, then the results of task 7.3 and calculation B would show that the Hopf link and the unlink with two components are definitely different links. Unfortunately, we are still left with the problematic Reidemeister move I. Before attending to this difficulty, it is worth practising this new type of *knot algebra* by doing one more calculation.

✻ **Task 7.8**   Determine the bracket polynomial of the trefoil in figure 7.4a. Be careful to proceed with great care, noting down each intermediate step, and remember that each pair of 'pointy' brackets should contain a small diagram.

Also determine the bracket polynomial of the mirror image of the trefoil in figure 7.4b. You may be able to use some of the steps from your previous calculation.

(a)   (b)

The trefoil and its mirror image.

— *Figure 7.4* —

Let us now consider what happens under a Reidemeister move of type I. Why did we delay looking at this move for so long? Calculation C will repay careful study.

Calculation C

What happens to the bracket polynomial under a Reidemeister move of type I?

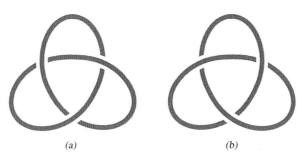

$$\left\langle \vphantom{x} \right\rangle \equiv x \left\langle \vphantom{x} \right\rangle + x^{-1} \left\langle \vphantom{x} \right\rangle$$

$$\equiv x\left[-x^2 - x^{-2}\right]\left\langle \vphantom{x} \right\rangle + x^{-1}\left\langle \vphantom{x} \right\rangle$$

$$\equiv -x^3 \left\langle \vphantom{x} \right\rangle.$$

Task 7.9   Calculation C shows that the bracket polynomial changes under one version of an 'untwist', one variation of a Reidemeister move of type I. Determine what happens if you perform the other 'untwist', a Reidemeister move of type I in the other direction (see figure 2.12a on page 14).

We conclude from calculation C and task 7.9 that the bracket polynomial is unfortunately not a true knot invariant: it fails for Reidemeister moves of type I.

Task 7.10   Find the bracket polynomial 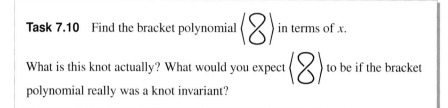 in terms of x.

What is this knot actually? What would you expect to be if the bracket polynomial really was a knot invariant?

In the next section we will show how to overcome the problem caused by Reidemeister moves of type I. Note that the simple expedient of making $-x^{-3} = 1$, by setting $x = -1$, say, does not work: not only would *all* the resulting polynomials be constant, but the Hopf link and unlink with two components, for example, would have the same polynomial, namely the constant polynomial -2.

7.2   The writhe

We almost discovered a knot invariant in the previous section, but it failed for moves of type I. We met something similar in chapter 6: the linking number is invariant under a Reidemeister move of type II or III, whereas a move of type I is irrelevant. A type I move makes no contribution to the linking number since it does not involve a crossing between different components.

We now define a new number, the *writhe* of a knot or link, denoted by $\omega(L)$. Section 6.3 on page 65 showed how to associate a sign ($+1$ or -1) to an oriented crossing. The writhe is calculated by associating a sign in this way to *every* crossing of an oriented knot or link and adding them all together.

Though the calculation of the writhe is very similar to that of the linking number, there are two important differences: firstly, every crossing is included, not just those between different components; secondly, there is no division by two.

Task 7.11   Determine the writhe of the oriented link in figure 7.5 on the facing page.

Task 7.12   Choose an orientation for the trefoil in figure 7.4a on the previous page and determine the writhe. Now determine the writhe if you had chosen the other orientation.

What happens if you consider the mirror image of the trefoil (figure 7.4b)? Calculate the writhe for this knot too.

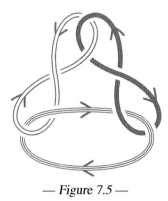

— Figure 7.5 —

Task 7.13 In task 7.12 you saw that the writhe does not change if you change the orientation of the trefoil knot. Is this only true for the trefoil or is it a general fact, true for all knots? Can you explain why? And what happens for links?

Task 7.14 Determine the writhe of the knot in figure 7.6.

— Figure 7.6 —

Task 7.15 Show that the writhe is invariant under Reidemeister moves of type II and III.

Task 7.16 What happens to the writhe if you perform a Reidemeister move of type I?

We conclude from tasks 7.15 and 7.16 that the writhe is invariant under a Reidemeister move of types II and III, but not under a move of type I, where the writhe changes by $+1$ or -1.

7.3 The *X*-polynomial

Neither the bracket polynomial nor the writhe is a true knot invariant: they both fail for Reidemeister moves of type I. In the polynomial an unwanted term $-x^3$ or $-x^{-3}$ crops up, and the writhe changes by $+1$ or -1. It was Jones' idea to combine both problems in an astute way. First of all he defined the *X-polynomial* of a knot or link L by

$$X(L) \equiv \left(-x^3\right)^{-\omega(L)} \langle L \rangle.$$

Now because the bracket polynomial $\langle L \rangle$ and the writhe $\omega(L)$ are both invariant under Reidemeister moves of type II and type III, so is the X-polynomial. But what about moves of type I?

In order to answer this question we consider two oriented knots (or links) K and L that are identical except for the region circled in figure 7.7, where we perform an 'untwist'.

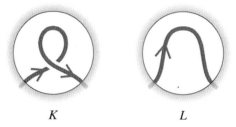

$$K \qquad\qquad\qquad L$$

Perform an 'untwist' on K to obtain L.

— *Figure 7.7* —

Now from calculation C on page 83 we know that $\langle K \rangle \equiv (-x^3)\langle L \rangle$, and the writhes of the two knots are related by $\omega(K) = \omega(L) + 1$. Therefore

$$X(K) \equiv \left(-x^3\right)^{-\omega(K)} \langle K \rangle$$

$$\equiv \left(-x^3\right)^{-(\omega(L)+1)} \left(-x^3\right)\langle L \rangle$$

$$\equiv \left(-x^3\right)^{-\omega(L)} \langle L \rangle$$

$$\equiv X(L).$$

What has happened here? Well, the writhe of K is one more than that of L and the bracket polynomial of K has an extra factor of $-x^3$. But the way Jones defined the X-polynomial means that these two effects cancel one another out. The result is that the X-polynomial is invariant under an 'untwist'.

> **Task 7.17**   Use the results of calculation C and task 7.16 to show that the X-polynomial is also invariant under a 'twist', the Reidemeister move of type I in the other direction.

We have now proved that *the X-polynomial is a true invariant for arbitrary oriented links*. For knots we have an even stronger statement: we can omit the adjective 'oriented' since we learned in task 7.13 that a change of orientation does not change the writhe, and the bracket polynomial does not depend on the orientation.

As an example, let us calculate the X-polynomial of the unlink with two components. We already know the bracket polynomial of this link from task 7.3 on page 81, namely $-x^2 - x^{-2}$. Since the two loops do not touch each other the writhe is simply zero, and $\left(-x^3\right)^{-0} \equiv 1$. Therefore

$$X\left(\bigcirc\bigcirc\right) \equiv -x^2 - x^{-2}.$$

> **Task 7.18** Determine the X-polynomials of the trefoil (figure 7.4a on page 83) and its mirror image (figure 7.4b).

Task 7.18 finally shows that the trefoil and its mirror image are different knots, something that we have been unable to do until now.

> **Task 7.19** Determine the X-polynomial of the Hopf link (figure 7.3 on page 82).
>
> What happens when you reverse the orientation of one of the components? And what happens if you do this for both components?

7.4 The Jones polynomial

We are finally in a position to define the Jones polynomial. The Jones polynomial $V(L)$ of a knot or link L is obtained from $X(L)$ by replacing the variable x by $t^{-\frac{1}{4}}$. In other words,

$$V(L) = X(L)\Big|_{x=t^{-\frac{1}{4}}}.$$

When the knot or link in question is clear the notation $V(t)$ is also used.

Now replacing x by $t^{-\frac{1}{4}}$ is likely to create an expression with rational powers of t, which would once again require the concept of a polynomial to be generalised. Surprisingly, however, it turns out that the Jones polynomial of a knot always has only integer powers (the proof of this fact is beyond the scope of this book). In a sense this fact provides the motivation for choosing to replace x by $t^{-\frac{1}{4}}$. It also provides a simple check of your working, which you may find useful in the remaining tasks in this section: the X-polynomial of a knot should only contain powers that are multiples of 4.

The situation for links is a little more complicated since the nature of the Jones polynomial for a link depends on the number of components: links with an odd number of components are like knots, the Jones polynomial has only integer powers; but the Jones polynomial of a link with an odd number of components has an extra factor $t^{\frac{1}{2}}$.

For example, using the result of task 7.19, we may find the Jones polynomial of the Hopf link in figure 7.3 on page 82:

$$V(\text{Hopf link}) \equiv -t^{\frac{1}{2}} - t^{\frac{5}{2}}.$$

Task 7.20 Determine the Jones polynomial of the trefoil and of its mirror image (see figure 7.4 on page 83).

Task 7.21 Determine the Jones polynomial of the figure of eight knot (see figure 3.10 on page 25) and of its mirror image.

We see that the Jones polynomial helps us to answer some of our previously unanswered questions, such as, can we distinguish a knot from its mirror image? In many cases the answer is yes, but there are cases where the Jones polynomial fails: it cannot distinguish the mirror images shown in figure 7.8, for example.

Mirror images with the same Jones polynomial.

— *Figure 7.8* —

The Jones polynomial also deals with another of our outstanding questions, concerning the number of knots with crossing number 6, which three-colourability could not answer (see task 5.11 on page 56).

✻ Task 7.22 Use the Jones polynomial to show that there are indeed three different knots with crossing number 6.

If two knots have different Jones polynomials, we can be sure that the knots are different too. Is the converse true—when two knots are different are their Jones polynomials also different? We mentioned above that this is not necessarily true for mirror images, but what about more general knots? Unfortunately, there exist distinct knots (and even knots having different crossing numbers, such as those in figure 7.9 on the next page) that have the same Jones polynomial.

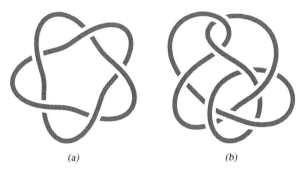

(a)   (b)

Two knots with the same Jones polynomial.

— Figure 7.9 —

It is not even known (at the time of writing) whether there exists a non-trivial knot whose Jones polynomial is identical to 1. As we know by now, the fact that nobody has found such a knot is not a proof that it does not exist.

However, all prime knots with 9 or fewer crossings do have distinct Jones polynomials. And as we have seen, in many cases the Jones polynomial can distinguish a knot from its mirror image. The Jones polynomial is a powerful tool, even though it is not perfect.

Is this the end of knot theory? No, far from it! The Jones polynomial was only the start of a new development in this branch of mathematics. The Jones polynomial was the first invariant in a list of new, stronger 'super-invariants'. Perhaps the most well-known was discovered independently and more or less simultaneously (around 1985) by several groups of people, and its name *HOMFLY polynomial* uses all the initials of its discoverers: Hoste, Ocneanu, Millett, Freyd, Lickorish, and Yetter. The polynomial has two variables and is a generalisation of the Jones polynomial. Though it is a stronger invariant than the Jones polynomial, the HOMFLY polynomial is still not perfect—there are some different knots which it cannot distinguish (including those in figure 7.9), and it fails to distinguish some mirror images (for example, those in figure 7.8 on the facing page).

Some 'perfect' knot invariants are known, too complicated for us to describe here, but they are difficult to calculate and so not useful in practice. What would be most welcome is an invariant which is easy to compute and which is able to distinguish many more knots. So the search continues!

.

POSTLUDE

A special trefoil

In the rest of this book we have considered knots as if they were made out of rope. To end the book, we look at a knot which has a special property when made from a rigid material such as wire.

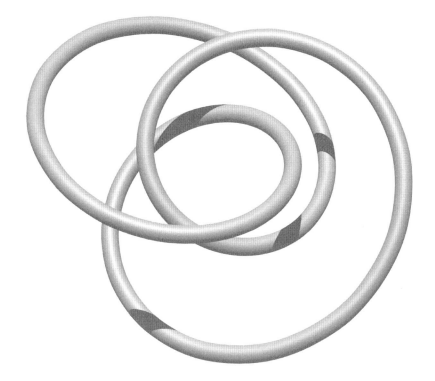

The knot shown above is just a trefoil knot, but its precise form makes it special.

> **Activity 1** Use a piece of rope to make the knot shown. Confirm that the knot is a trefoil.

Suppose you made the knot out of rigid wire rather than rope, and placed it on a table. How many times would the knot touch the table? Would the number of points of contact change if you put the knot down on the table in some other way?

You probably expect that any wire knot resting on a table will touch the table in three points, or more. After further thought you might anticipate that occasionally the knot may touch the table in only two points, just as the runners of a rocking chair touch the floor in only two points, but then by rocking the knot you will eventually bring more than two points into contact.

In 1980 the American mathematician Michael Freedman[*] asked, essentially, whether there was an example of a knot which only ever touched the table twice. In other words, rocking the knot never results in a situation where three (or more) points are in contact with the table at the same time.

Since we are doing mathematics and not actually dealing with real objects we make some assumptions: firstly we assume that the table is perfectly flat; secondly, as usual, we assume that the knot is a curve with no thickness, unlike the thickened version shown in the figure.

The knot shown overleaf was discovered in 1990 by Hugh Morton of Liverpool University in answer to Freedman's question [31]. This knot will never have more than two points in contact with the table—attempts to rock it far enough so that a third point comes into contact with the table fail because one of the two pieces of the curve disappears "inside". The knot not only has the property that it will only ever touch the table twice, but Morton also proved that you cannot even insert a sheet of glass which touches the knot in three points.

In fact Morton described a whole family of such knots[†]. All of them, like the one shown, are $(2, 3)$-torus knots (see page 73). So all of them are trefoils. What gives them their special property is the way they are constructed.

---

[*]Freedman won the Fields medal in 1986 for his work on the Poincaré conjecture.
[†]For the curious, the curves have the parametric equations

$$x = \frac{a\cos 3t}{1 - b\sin 2t}, \qquad y = \frac{a\sin 3t}{1 - b\sin 2t}, \qquad z = \frac{b\cos 2t}{1 - b\sin 2t},$$

where $a^2 + b^2 = 1$ and $a, b \neq 0$.

Solutions

Tasks

2.1 (a) The *bowline* is one of the most common ways to make a fixed loop; it is quick to tie and untie, and does not slip or jam. The name is pronounced '*bow*-linn'.

   (b) The *reef knot*, or *square knot*, is used to secure a rope around an object, and is therefore a *binding knot*. Though commonly used to tie two ropes together, that is, as a *bend*, this is *not recommended* since the result is an unstable knot.

2.5 Diagrams (a) and (b) show actual knots; both are the same knot. Diagram (c) shows an unknotted loop.

2.6 Both knots can be deformed into a circular ring.

2.7 Diagrams (a) and (b) show the unknot, whereas diagram (c) does not.

2.8 The knots are different since diagram (b) shows the unknot, whereas diagram (a) does not.

2.9 (a) Type II, 'unpoke': 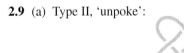   (b) Type III, 'slide':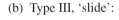

   (c) Type I, 'untwist':

2.10 One method is to use three 'unpokes' followed by an 'untwist'. No method can use fewer than four moves—to see why, consider what happens to the number of crossings for each type of move.

2.11 Only two moves are possible: a 'twist' (type I) or a 'poke' (type II). Note that you need two consecutive over- or under-crossings in order to perform a move of type III, which is not the case here.

3.1 (a) 3   (b) 4

The knots can be transformed into one another. For example, consider the right-hand knot and the crossing at the bottom left. Moving the upper strand of this crossing to the right "untwists" the crossing at the centre.

3.2 (a) 0   (b) 3   (c) 4   (d) 3   (e) 0   (f) 24

3.3 There are no knots with crossing number 1 or 2.

3.4 It is actually easier initially to ignore the nature of the crossings—whether they are 'over-' or 'under-crossings'. In other words, just consider possible *projections* with two crossings. (The case for one crossing can be dealt with in a similar way.)

One way to proceed is as follows. At least one strand connects the two crossings, therefore at least two strands do—can you see why? Consider these two strands, which together essentially form a loop of rope, and determine the possible arrangements for the crossings and the other strands.

There are exactly five possible projections with two crossings:

Now note that the nature of the crossings is irrelevant since all cases lead to the unknot.

3.5 The crossing number is 3; the knots are the same (see task 3.6).

3.6 Starting with the overhand knot, take hold of the middle of the strand at the very top of the diagram; move your hand down to the left, bringing the loop across the top of the knot; then rearrange the loops a little and you are done.

3.8 (a) the unknot   (b) the trefoil   (c) the figure of eight   (d) the trefoil
(e) the unknot

3.10 The knots are different.

3.11 The answers depend on the labelling and the starting point, but possible words are:
(a) ABCDEABCDE;   (b) ABCDEABEDC or ABCDECBADE.

3.12 One possible rule is 'repeat the first two letters in the same order, then repeat the next three letters in reverse order'.

3.13 (a) 3-twist   (b) cinquefoil   (c) cinquefoil   (d) 3-twist

4.1 It is not possible to deform the trefoil into its mirror image.

4.2 (a) trefoil   (b) trefoil   (c) mirror image

4.3 It is possible to deform the figure of eight into its mirror image.

4.4 By applying two 'pokes' and a 'slide' we can slide a strand from one side of a crossing to the other:

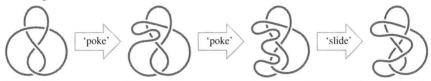

We shall want to use this procedure again, so let us use the term 'p-p-s' for the sequence of moves 'poke'-'poke'-'slide' applied in this way. In other words, 'p-p-s' moves a strand across a crossing.

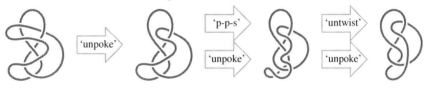

4.5 Both the cinquefoil and the 3-twist are chiral.

4.6 Rotating the left-hand diagram through 90° clockwise gives the right-hand diagram, so the diagrams show the same knot. But the diagrams also show mirror images, so the knot is amphichiral.

The knot is the figure of eight.

4.7 Rotating the diagram through 180° gives a diagram of the mirror image. Therefore the knot is amphichiral.

4.8 It is clear that both methods give the same knot *projection*. The key difference is what happens to crossings.

When the knot is lifted and turned over, an overstrand becomes an understrand, and *vice versa*.

However, when the diagram on the sheet is turned over, solid and broken lines do not change, so that an overstrand remains an overstrand and an understrand remains an understrand. This is the opposite of what happens when the knot is turned over, so the effect is the same as reflecting the original knot diagram, reversing all the crossings.

4.9 Diagram (a) shows the reef knot (see the solution to task 2.1). Diagram (b) shows the *granny knot*, another binding knot like the reef knot.

Each knot is a combination of two trefoils. The reef knot is amphichiral; the granny knot is chiral.

4.10 The composition of a knot K and the unknot is always K.

4.11 The procedure gives two ways of combining two knots, because one of the knots may be turned over before the ends are connected. Of course, this does not tell us whether the results are different knots.

4.12 The knot is a composition of the trefoil and the figure of eight.

4.13 Two of the crossings in (a) can be removed and the knot is a composition of two figures of eight. The knot in (b) is a prime knot with crossing number 10.

4.14 Suppose that a given knot could be unknotted by composing it with another knot. Then we would have a composition of two knots that was the unknot, which contradicts the assumption that the unknot is not composite. Hence the supposition is wrong, so it is impossible to undo a given knot by composing it with another knot.

4.15 Only (e) and (f) are prime, the rest are composite: (a), (b) and (d) are compositions of two trefoils; (c) is a composition of a trefoil and the figure of eight; (g) is a 'triple' composition of two trefoils and a figure of eight.

4.16 There are no knots with crossing number 1 or 2, so that to obtain crossing number 7 we have to compose knots with crossing numbers 3 and 4. (Since we are dealing with small crossing numbers we may assume that no crossings can disappear in the process, as stated in the text.) However, there is only one knot with crossing number 3 (the trefoil) and one knot with crossing number 4 (the figure of eight). These two knots may be composed in the two ways shown below.

This shows there are no more than two knots of this form, but does *not* show that these two are in fact different knots (which they are).

4.17 Since we are dealing with small crossing numbers we may assume that no crossings can disappear in the process, as stated in the text. There are no knots with crossing number 1 or 2, so that the only way to obtain crossing number 10 is to compose knots with crossing numbers 5 and 5, or 6 and 4, or 7 and 3, or 4, 3 and 3.

4.18 Changing one crossing results in the unknot.

4.19 (b) The top left crossing has been changed; the result is the unknot.

    (c) All four crossings have been changed; the result is the mirror image, which is also a figure of eight.

4.20 One way is to change the crossings indicated:

(a)   (b)

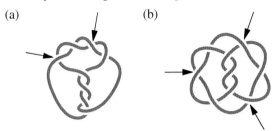

4.22 The unknotting number of the 8-twist is 1. For example, change the top crossing.

4.23 The unknotting number of the cinquefoil is 2; that of the 3-twist is 1.

5.1 Only two colours come together at the lower two crossings, so the second rule is broken.

5.2 The real reason is that the knot in figure 5.1b is the unknot, which is not three-colourable. If you try to colour the knot, you will find that the first colour you use will appear twice at some crossing and therefore the other strand at that crossing has to have the same colour. Continuing in this way results in all the strands being the same colour, contrary to what is required.

5.3 The trefoil is three-colourable (see figure) and the unknot is not. We shall see later that the figure of eight is also not three-colourable.

5.4 (a) Both knots are three-colourable, so that three-colourability does not distinguish these two knots.

    (b) The left-hand knot is not three-colourable and the right-hand knot is, so that three-colourability distinguishes these two knots.

5.5 Whichever variation of a type I Reidemeister move you do, the two strands leaving the circle will have the same colour. This means, first of all, that at least one of the other colours appears somewhere outside. It also means that anything inside the circle can be left in this one colour. Therefore the rules are preserved in all cases.

5.6 For the 'unpoke' just reverse the chain of arguments for the 'poke'.

5.7 There are three strands, which we refer to as the 'top', 'middle' and 'bottom' strands (taking the diagram to be looking down from above). Consider the two crossings, between the top and middle strands, and between the middle and bottom strands. According to the rules, each crossing involves either just one colour or all three colours, so altogether there are four possibilities; in each case there is no choice for the colours at the third crossing, these are determined by the rules.

The following diagrams show the four possibilities, and how three-colourability is maintained, for one version of the Reidemeister move. For the other version the diagrams are just mirror images.

5.8 Use the method of the text: start at a crossing and follow through each possibility. You may use one or three colours at the starting crossing, but in both cases you will find there is a problem: either colours clash, or only one colour is used.

5.9 Since the trefoil is three-colourable and the figure of eight is not the two knots are different.

5.10 (a)   (b)   (c)

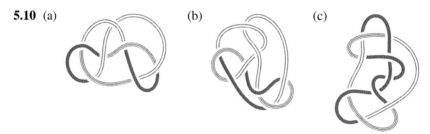

5.11 The knot 6_1 is three-colourable, but 6_2 and 6_3 are not. Therefore 6_1 is definitely different to the other two knots, but three-colourability is unable to distinguish between 6_2 and 6_3.

5.12 Either at least two of the knots will be three-colourable, or at least two will not be three-colourable. In both cases, it is not possible to distinguish at least two of the knots.

5.13 The knots shown in (b), (d), (e) and (f) are three-colourable; (a) and (c) are not.

6.1 Only figure 6.2b shows a knot; all the others are links, as shown:

6.2 The following figures show how to deform one link to the other. We leave it to you to describe which Reidemeister moves are needed at every stage. To help you to check, in total you should find that you need one move of type I, five of type II, and eight of type III.

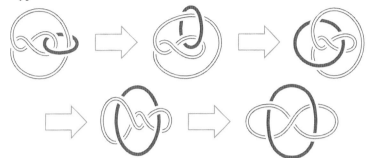

6.3 One possible sequence of moves is shown in the following figures.

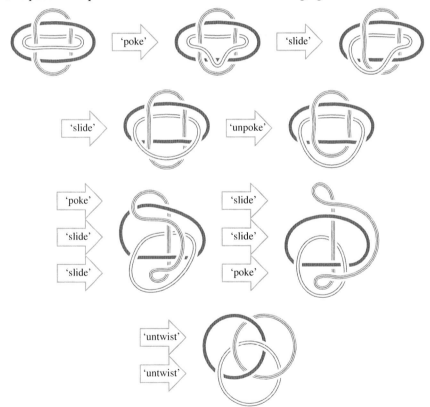

6.4 (a) 3   (b) 2   (c) 4

6.6 The strand of one loop passes through the other loop twice, so the link may be considered to be linked twice. If the definition of linking number did not include the phrase 'divide by 2' then a value of 4 would be obtained, rather than the expected value of 2.

6.7 The linking number of a knot is 0, because at every crossing both strands belong to the same component, so none of the crossings is labelled.

6.8 The linking number is either +3 or −3. By reversing the orientation of just one component you change every +1 crossing to a −1 crossing and *vice versa* (see figure 6.10). Hence the sign of the linking number is changed. But changing the orientation of both components leaves the linking number unchanged.

6.9 If you perform a Reidemeister move of type II, then either two new crossings are created (a 'poke'), or two are annihilated (an 'unpoke'). However, the two crossings always have opposite signs (one +1 and one −1) so that in either case the total does not change. Hence the move does not change the linking number.

If the orientation of one of the components is reversed, then the signs of all the crossings are changed so the linking number is multiplied by −1, both before and after the move.

We conclude that the linking number does not change under a Reidemeister move of type II.

6.10 Observe that under a 'slide' all that changes are the relative positions of the three crossings. For each crossing, the directions and orientations of the strands do not change; and clearly which component a strand belongs to remains the same. Hence the calculations for the linking number are the same for each diagram.

6.11 The unlink with two components has linking number 0, whereas the Hopf link has linking number 1, hence the two links are not equivalent.

6.12 The linking numbers are 3 for (a) and 2 for (b), so the links are different.

6.13 Both the Whitehead link and the unlink with two components have linking number 0.

6.14 The linking number is 3. (Note that the linking number does not distinguish this link from the one shown in figure 6.12. It can also be shown that both these links have crossing number 9.)

6.15 For the unlink with two components, just use two different colours for the two components. For the unlink with $n > 2$ components, do the same for the first two components and use any colour for the remaining components.

6.16 Label the sections of the strands of the link s, t, u, v and w, as shown.
Now if s and t have the same colour then the whole link has the same colour, which is not allowed. Hence s and t have different colours. It follows that u and v have the third colour. But now it is impossible to choose a colour for w.

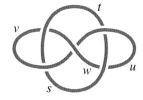

Therefore the Whitehead link is not three-colourable and hence is definitely not equivalent to the unlink with two components, which *is* three-colourable.

6.17 Both the link in figure 6.18 and the Whitehead link have two components. Both also have linking number 0. However, the Whitehead link is not three-colourable, whereas the link in figure 6.18 is three-colourable, as shown.

6.18 Each link has two components, crossing number 9 and linking number 3, but the left-hand link is three-colourable, as shown, whereas the other is not.

7.1 $\langle \bigcirc \sqcup \bigcirc \rangle \equiv z \langle \bigcirc \rangle$　(from Rule 3)

$\equiv z \times 1$　(from Rule 1)

$\equiv z.$

7.2 $\left\langle \vphantom{} \right\rangle \equiv x \left\langle \vphantom{} \right\rangle + y \left\langle \vphantom{} \right\rangle$　(from Rule 2)

$\equiv x \times 1 + y \times z$　(from Rule 1 and task 7.1)

$\equiv x + yz.$

7.3 $\langle \bigcirc \sqcup \bigcirc \rangle \equiv -x^2 - x^{-2}.$

7.4 $\left\langle \vphantom{} \right\rangle \equiv x \left\langle \vphantom{} \right\rangle + x^{-1} \left\langle \vphantom{} \right\rangle$　(from Rule 2)

$\equiv x \left\langle \vphantom{} \right\rangle + x^{-1} \left\langle \vphantom{} \right\rangle$　(making two moves of type II)

$\equiv \left\langle \vphantom{} \right\rangle.$　(from Rule 2)

7.5 When L is the unlink with 3 components, $\langle L \rangle \equiv \left(-x^2 - x^{-2} \right)^2.$

When L is the unlink with n components, $\langle L \rangle \equiv \left(-x^2 - x^{-2} \right)^{n-1}.$

7.6 Line 1: Rule 2a.

Line 2: Rule 2a (twice).

Line 3: Rules 3, 1, 1 and 3.

7.7 In the calculation for the mirror image of the Hopf link, compared with calculation B, you need to use Rule 2 instead of Rule 2a, or *vice versa*, with the effect that x and

x^{-1} are interchanged. Apart from this, the calculation is the same. However, the final polynomial is symmetric with respect to positive and negative powers, so is unchanged.

7.8 We make use of the fact, established in calculation B on page 82, that the bracket polynomial of the Hopf link is $-x^4 - x^{-4}$.

$$\langle \text{⬡} \rangle \equiv x \langle \text{⬡} \rangle + x^{-1} \langle \text{⬡} \rangle \qquad \text{(from Rule 2)}$$

$$\equiv x^2 \langle \text{⬡} \rangle + \langle \text{⬡} \rangle + x^{-1}\left[-x^4 - x^{-4}\right] \qquad \text{(from Rule 2a)}$$

$$\equiv x^2\left[-x^2 - x^{-2}\right]\langle \text{⬡} \rangle + \langle \text{⬡} \rangle - x^3 - x^{-5} \qquad \text{(from Rule 3)}$$

$$\equiv -x^4 \langle \text{⬡} \rangle - x^3 - x^{-5}$$

$$\equiv -x^5 \langle \text{⬡} \rangle - x^3 \langle \text{⬡} \rangle - x^3 - x^{-5} \qquad \text{(from Rule 2a)}$$

$$\equiv -x^5\left[-x^2 - x^{-2}\right] - x^3 - x^3 - x^{-5} \qquad \text{(from Rules 3 and 1)}$$

$$\equiv x^7 - x^3 - x^{-5}.$$

When calculating the bracket polynomial for the mirror image of the trefoil, we may use the same steps as above, except that we need to use Rule 2a instead of Rule 2, and *vice versa*. Going through the steps carefully, you will find that the final result is $x^{-7} - x^{-3} - x^5$.

7.9
$$\langle \text{⬡} \rangle \equiv x \langle \text{⬡} \rangle + x^{-1} \langle \text{⬡} \rangle$$

$$\equiv x \langle \text{⬡} \rangle + x^{-1}\left[-x^2 - x^{-2}\right]\langle \text{⬡} \rangle$$

$$\equiv -x^{-3} \langle \text{⬡} \rangle.$$

7.10
$$\langle \text{⬡} \rangle \equiv x \langle \text{⬡} \rangle + x^{-1} \langle \text{⬡} \rangle \qquad \text{(from Rule 2)}$$

$$\equiv x \times 1 + x^{-1}\left[-x^2 - x^{-2}\right] \qquad \text{(from Rule 1 and task 7.3)}$$

$$\equiv -x^{-3}.$$

Since the knot is just an unknot we would expect the polynomial to be 1.

7.11 Six crossings have sign -1 and two have sign $+1$, so the writhe is -4.

7.12 Whatever the orientation, the writhe of the trefoil in figure 7.4a is −3. The writhe of the mirror image is +3.

7.13 For knots, the orientation is irrelevant to the writhe—if you change the orientation, then at every crossing both arrows are reversed, resulting in the same sign for the crossing.

However, for links this is no longer true. From the argument above, only if you change the orientation of every component will the writhe remain unchanged. If you reverse the orientation of some but not all of the components, then some crossings will change sign and some will not, depending on whether both strands are changed, or only one is, or neither is.

Hence the writhe is only well-defined for oriented links.

7.14 Three crossings have sign −1 and three have sign +1, so the writhe is 0.

7.15 We only consider that part of the knot where the Reidemeister move is performed: since the signs of any other crossings remain the same the writhe is not affected by them.

For a move of type II, whatever the orientation of the strands:

 has no crossings;

 has one crossing with sign +1 and one with sign −1.

In each case, in total nothing is contributed to the writhe, so that the writhe is invariant under a move of type II.

Under a move of type III, a 'slide', all that changes are the relative positions of the three crossings. For each crossing, the directions and orientations of the strands do not change. Hence the signs of the three crossings do not change, so the crossings contribute the same total to the writhe in each diagram. Therefore the writhe is invariant under a move of type III.

7.16 The writhe is *not* invariant under a Reidemeister move of type I, because no matter what orientation you choose:

 has no crossings, so contributes nothing to the writhe;

 has one crossing with sign +1, so contributes +1 to the writhe;

 has one crossing with sign −1, so contributes −1 to the writhe.

7.17 $X(L) \equiv \left(-x^3\right)^{-\omega(L)} \langle L \rangle$

$$\equiv \left(-x^3\right)^{-(\omega(K)-1)} \left(-x^{-3}\right)\langle K \rangle$$

$$\equiv \left(-x^3\right)^{-\omega(K)} \langle K \rangle$$

$$\equiv X(K).$$

7.18 We put together information we have already found: from task 7.12 we know that the writhe of the trefoil is -3; and from task 7.8 we know that the bracket polynomial is $x^7 - x^3 - x^{-5}$. So the X-polynomial of the trefoil is

$$\left(-x^3\right)^3 \times \left(x^7 - x^3 - x^{-5}\right) = -x^{16} + x^{12} + x^4.$$

Similarly the X-polynomial of the mirror image is

$$\left(-x^3\right)^{-3} \times \left(x^{-7} - x^{-3} - x^5\right) = -x^{-16} + x^{-12} + x^{-4}.$$

7.19 Depending on the choice of orientations, the writhe of the Hopf link is either $+2$ or -2. The bracket polynomial of the Hopf link was found in calculation B on page 82. Therefore the X-polynomial is either $-x^{-2} - x^{-10}$ or $-x^2 - x^{10}$.

7.20 The Jones polynomial of the trefoil is $-t^{-4} + t^{-3} + t^{-1}$ and the mirror image has Jones polynomial $t + t^3 - t^4$.

7.21 The Jones polynomial of the figure of eight is $t^{-2} - t^{-1} + 1 - t + t^2$. The mirror image has the same Jones polynomial.

7.22 The following table gives the Jones polynomials of the three knots, using the notation given on page 111.

Knot	Jones polynomial
6_1	$t^{-4} - t^{-3} + t^{-2} - 2t^{-1} + 2 - t + t^2$
6_2	$t^{-5} - 2t^{-4} + 2t^{-3} - 2t^{-2} + 2t^{-1} - 1 + t$
6_3	$-t^{-3} + 2t^{-2} - 2t^{-1} + 3 - 2t + 2t^2 - t^3$

Since the three polynomials are different, we know that the three knots are different.

Activities

Knots in paper

1 It is possible to *prove* that the final figure is a regular pentagon using only the facts that the sides of the strip are parallel and the strip has a constant width. For example, these facts are sufficient to prove that the shaded shape in the figure alongside is a rhombus, which shows that two sides of the pentagon are equal.

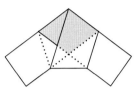

2 Cutting the Möbius band in 'thirds' creates two interlinked bands, an example of a link—see chapter 6. One of the bands is another Möbius band, therefore one-sided; the other band is formed by giving a strip two half-twists, which means that it is two-sided.

3 The band is another version of the Möbius band, still one-sided. When cut in half the result is a single band in the form of a trefoil knot—introduced on page 23.

4 Cutting the Möbius shorts 'in half' creates a rectangular frame, as shown, that is, the unknot—defined on page 12. The cut forms the outer edge of the frame.

Celtic knots

1 From left to right 5 and 12 crossings need to be removed, as highlighted below.

2 The 3×5 plaited rectangle has 1 strand; the 3×9 plaited rectangle has 3 strands.
The Celtic knot formed from the 3×5 plaited rectangle has 3 strands.
The two Celtic knots in activity 1 (from left to right) have 2 and 5 strands.
The left-hand Celtic knot on page 17 has 2 strands; the right-hand knot has 4 strands.

3 The highest common factor of 4 and 10 is 2, so a 4×10 plaited rectangle has 2 strands.

Tie knots

1 The Windsor knot forms the trefoil.

Torus knots

1 The $(3, 2)$-torus knot is the trefoil.

2 The $(3, 1)$-torus knot is the unknot.

3 The $(2, 3)$-torus knot is also the trefoil.

4 The $(4, 2)$-torus knot requires two pieces of rope; the result is the link 4_1^2 with two components, shown on page 114.

Bibliography

The lists below include recommendations for further reading as well as items of historical interest. The notes should help you to decide where to go next to find out more.

Books

[1]   Colin C. Adams.
      The Knot Book. An Elementary Introduction to the Mathematical Theory of Knots.
      American Mathematical Society, 2004. ISBN: 978-0-8218-3678-1.
      ↪ Covers more mathematics than *Knots Unravelled* at a higher level.

[2]   Meike Akveld. *Knoten in der Mathematik*. German. Orell Füssli, 2007.
      ISBN: 978-3-280-04050-8.

[3]   Clifford W. Ashley. *The Ashley Book of Knots*. Ed. by Geoffrey Budworth.
      Faber and Faber, Apr. 1993. ISBN: 978-0571096596.
      ↪ A classic reference work on 'real-world' knots.

[4]   David W. Farmer and Theodore B. Stanford.
      Knots and Surfaces: a guide to discovering mathematics. Vol. 6.
      Mathematical World. American Mathematical Society, 1996.
      ISBN: 978-0-8218-0451-3.

[5]   Thomas Fink and Yong Mao.
      The 85 Ways to Tie a Tie: The Science and Aesthetics of Tie Knots.
      Fourth Estate Ltd, Nov. 2001.

[6]   Heather McLeay. *The Knots Puzzle Book*. Key Curriculum Press, 2008.
      ISBN: 978-1-55953-000-2.

[7] Des Pawson. *Handbook of Knots*. Expanded edition. Dorling Kindersley, 2004.
ISBN: 978-1-4053-0467-2.
⌐ A modern introduction to rope knots.

[8] Burkard Polster. *The Shoelace Book. A Mathematical Guide to the Best (and Worst) Ways to Lace Your Shoes*. Mathematical World Volume 24.
American Mathematical Society, 2006. ISBN: 978-0-8218-3933-1.
⌐ All you ever wanted to know about tying shoelaces.

[9] Dale Rolfsen. *Knots and Links*. AMS Chelsea Publishing.
American Mathematical Society, 1976. ISBN: 978-0-8218-3436-7.
⌐ A classic: authors still refer to the Rolfsen table of knots.

[10] Phil D. Smith. *Knots for Mountaineering, Camping, Climbing, Utility, Rescue, Etc.*
Publisher unknown, c1960.

Online

[11] Dror Bar-Natan and Scott Morrison. *The Knot Atlas*. 2011.
URL: `http://katlas.org/`.
⌐ Aims to be a complete user-editable knot atlas.

[12] Ronald Brown. *Symbolic Sculptures by John Robinson*. 1993.
URL: `http://www.popmath.org.uk/sculpture/pages/bangor.html`.

[13] Jae Choon Cha and Charles Livingston. *KnotInfo: Table of Knot Invariants*. 2011.
URL: `http://www.indiana.edu/~knotinfo`.

[14] Peter R. Cromwell. *The Borromean Rings*. July 2007.
URL: `http://www.liv.ac.uk/~spmr02/rings/index.html`.
⌐ Includes some historical background and examples of Borromean symbolism.

[15] Alex Feingold. *Alex Feingold*. 2011.
URL: `http://www.math.binghamton.edu/alex/`.

[16] Helaman Ferguson. *Helaman Ferguson sculpture*. 2003.
URL: `http://www.helasculpt.com/`.

[17] Thomas Fink. *Encyclopedia of Tie Knots*.
URL: `http://www.tcm.phy.cam.ac.uk/~tmf20/tieknots.shtml`.

[18] Thomas Fink. *The 85 Ways to Tie a Tie*.
URL: `http://www.tcm.phy.cam.ac.uk/~tmf20/85ways.shtml`.

[19] Ortho Flint and Stuart Rankin. *Knotilus*. 2010.
URL: `http://knotilus.math.uwo.ca/`.
⌐ An interactive tool for drawing and studying the various properties of knots and links.

[20]   IMU. *New IMU Logo based on the tight Borromean rings*. 2006.
        URL: `http://torus.math.uiuc.edu/jms/Images/IMU-logo/`.

[21]   International Guild of Knot Tyers. "Knot Charts". 2011.
        URL: `http://www.igkt.net/pdf/KnotChartsWeb.pdf`.
        ⟳ Diagrams showing how to tie 'real-world' knots.

[22]   Ivars Peterson.
        Ivars Peterson's MathTrek: Sand Drawings and Mirror Curves. MAA Online.
        Mathematical Association of America. Sept. 2001.
        URL: `http://www.maa.org/mathland/mathtrek_9_24_01.html`.

[23]   Robert G. Scharein. *The KnotPlot Site*. 2011. URL: `http://knotplot.com/`.
        ⟳ A collection of knots and links, viewed from a (mostly) mathematical perspective.

[24]   Carlo H. Séquin. *Carlo H. Séquin*. 2011.
        URL: `http://www.cs.berkeley.edu/%7Esequin/`.

[25]   N. J. A. Sloane. *Sequence A002863. Number of prime knots with n crossings*.
        The On-Line Encyclopedia of Integer Sequences. 2010.
        URL: `http://oeis.org/A002863`.

[26]   *Swiss Knots 2011. Knot Theory and Algebra*. 2011. URL: `http:`
        `//www.math.uzh.ch/swissknots2011/index.php?swissknots2011`.

[27]   Wikipedia contributors. *Chinese knotting*. Wikipedia, The Free Encyclopedia.
        2011. URL: `http:`
        `//en.wikipedia.org/w/index.php?title=Chinese_knotting&oldid=4`
        `16406335`.

Published articles

[28]   J. W. Alexander and G. B. Briggs. "On types of knotted curves".
        In: *Annals of Mathematics* 28 (1926/7), pp. 562–586.

[29]   Ralph P. Boas Jr. "Möbius Shorts".
        In: *Mathematics Magazine* 68.2 (Apr. 1995), p. 127.

[30]   Gwen Fisher and Blake Mellor. "On the Topology of Celtic Knot Designs".
        In: *Proceedings of the 7th Annual International Conference of Bridges*. 2004.
        URL: `http://myweb.lmu.edu/bmellor/papers.html`.

[31]   H. R. Morton. "Trefoil knots without tritangent planes".
        In: *Bulletin of the London Mathematical Society* 23.100 (Jan. 1991), pp. 78–80.

[32]   H. Reidemeister. "Knoten und Gruppen".
        In: *Abh. Math. Sem., Univ. Hamburg* 5 (1926), pp. 7–23.

[33] De Witt Sumners.
"Lifting the Curtain: Using Topology to Probe the Hidden Action of Enzymes".
In: *Notices of the AMS* 42.5 (May 1995), pp. 528–537.
URL: http://www.ams.org/notices/199505/sumners.pdf.

[34] Sir W. H. Thomson. "On Vortex Motion".
In: *Transactions of the Royal Society of Edinburgh* XXV (1867), pp. 217–260.

Table of knots and links

The table includes diagrams of:

    ▷ all prime knots with crossing number less than 9; and

    ▷ all prime links with crossing number less than 7.

Mirror images are *not* included.

Key

X_n: the nth knot with crossing number X.

X_n^c: the nth link with crossing number X and c components.

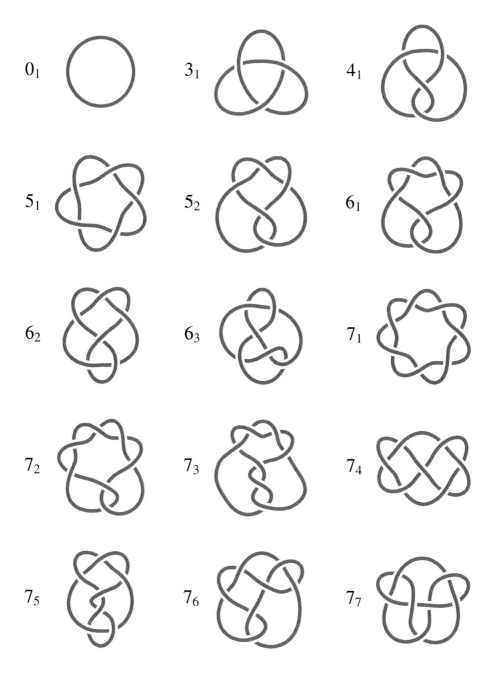

0_1 3_1 4_1

5_1 5_2 6_1

6_2 6_3 7_1

7_2 7_3 7_4

7_5 7_6 7_7

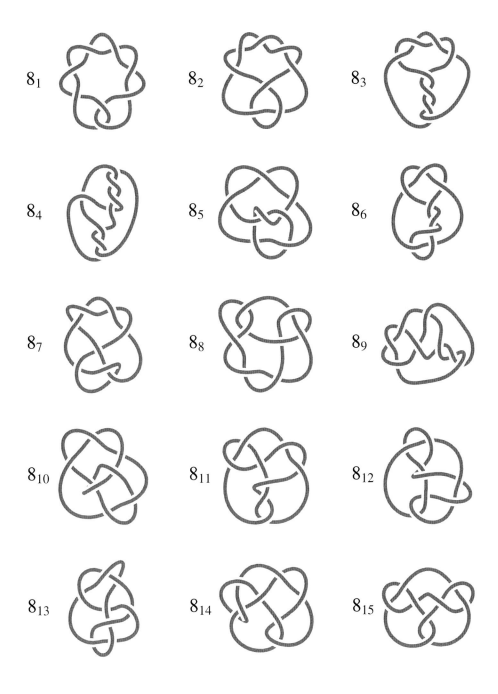

8_1 8_2 8_3

8_4 8_5 8_6

8_7 8_8 8_9

8_{10} 8_{11} 8_{12}

8_{13} 8_{14} 8_{15}

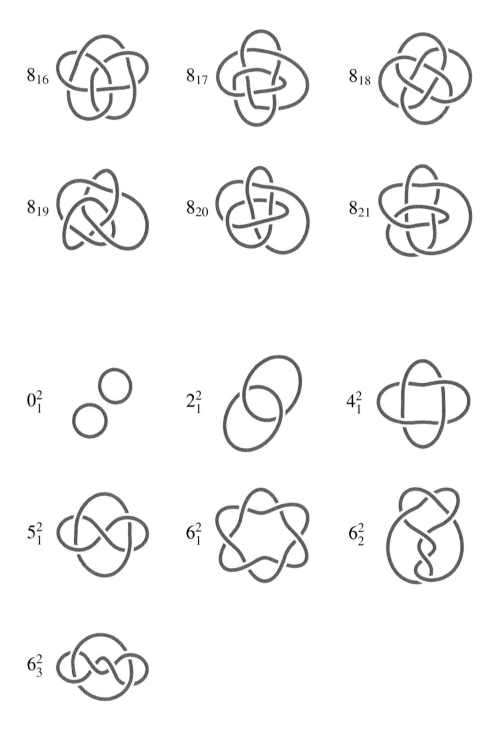

8_{16}

8_{17}

8_{18}

8_{19}

8_{20}

8_{21}

0_1^2

2_1^2

4_1^2

5_1^2

6_1^2

6_2^2

6_3^2

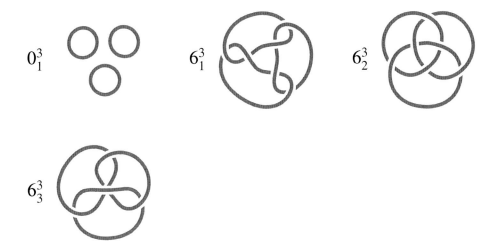

0^3_1

6^3_1

6^3_2

6^3_3

Glossary

For ease of reference, we have collected together the meanings of some of the terms used in knot theory. Please refer to the index for any term not included here.

amphichiral

A knot is amphichiral if it *is* equivalent to its mirror image. See page 35.

chiral

A knot is chiral if it is *not* equivalent to its mirror image. See page 34.

composite

A knot or link is composite if it is the composition of non-trivial links. See page 39.

crossing number

The crossing number of a knot is the *minimal* number of crossings among *all* possible knot diagrams of the knot. See page 21.

diagram

A knot diagram is a knot projection, where all the crossings are distinct and clear, with the crossings changed to indicate over- and under-passes. See page 11.

invariant

A knot invariant is a property of a knot which does not change under the three Reidemeister moves. See page 49.

non-trivial

A knot or link is non-trivial if it is definitely knotted or linked.

prime

A knot or link is prime if it is not the composition of non-trivial links. See page 62.

projection

A knot projection is the shadow of a knot on a plane (when the light source is an infinite distance away, so the light rays are parallel). See page 9.

trivial

A knot or link is trivial if it is neither knotted nor linked.

unknot

A knot is the unknot if it can be deformed into a circular ring. See page 12.

unlink

A link is the unlink if it can be deformed into unlinked circular rings. See page 65.

Index